Foetus into Man

*Physical Growth from
Conception to Maturity*

J. M. Tanner

SECOND EDITION

CASTLEMEAD PUBLICATIONS
WARE

First Edition Published in 1978

Second Edition Published in 1989

CASTLEMEAD PUBLICATIONS
Swains Mill, 4A Crane Mead
Ware, Herts, SG12 9PY

Publishing Division of
WARD'S PUBLISHING SERVICES

Hardback   ISBN 0 948555 23 8
Paperback ISBN 0 948555 24 6

© J. M. Tanner 1989

British Library Cataloguing in Publication Data
Tanner, J.M. (James Muorilyan)
   Foetus into man. -- 2nd ed.
   1. Man. Growth
   I. Title
   612.6

ISBN 0–948555–23–8
ISBN 0–948555–24–6 pbk

Set in 10/12 Times Roman
by Eurotype Ltd,, Hertford
Printed and bound in Great Britain
by Courier International Limited, Tiptree, Essex

Preface to the Second Edition

A good deal has happened since 1977 in the field of human auxology, the study of growth in its morphological and physiological aspects. I have rewritten this book accordingly, to bring it up to date in all respects, including Further Reading and References. The chapters which have been altered most are those on endocrinology (of course) and the interaction of heredity and environment in the control of growth.

Our understanding of the endocrine control of growth and particularly of the changes of puberty has advanced greatly in the last ten years, thanks above all to the elegant work of Ernst Knobil and his collaborators at Pittsburgh and elsewhere. This chapter has been entirely rewritten.

The chapter on factors affecting growth, in contrast, has chiefly been expanded. The basic situation has not changed greatly, but numerous studies of high quality have added details or made firm conclusions that were previously only tentative. I have particularly expanded the sections on the interaction of nutrition and infection on growth; on social class differences and the secular trend; and on growth as a measure, or proxy, of health, and thus growth of children as a mirror of the state of a society. This public health use of growth, or auxological epidemiology, has come increasingly to the fore, especially with the establishment of a national system of continuous growth surveillance in the United Kingdom.

As a result of these revisions, the book now goes a little deeper into these particular aspects of auxology than did the first edition, but without, I hope, impairing its accessibility to the teacher or parent innocent of biological or medical training. I trust that they, as well as medical students, nutritionists, and paramedics of all kinds who deal with children, will find in these pages the basis of an attitude toward human biology that will serve them well.

As before, I have to thank many friends, colleagues and students for criticisms of the draft, and Jan Baines Preece for assembling and processing the new text.

J M TANNER
London and Houston
March 1989

v

Contents

Preface to the Second Edition   *Page* v

Introduction   1

1   The Curve of Growth   6

2   Cells and the Growth of Tissues   24

3   Growth Before Birth   36

4   Sex Differentiation up to Puberty   51

5   Puberty   58

6   Developmental Age, and the Problems of Early and Late
    Maturers   75

7   The Endocrinology of Growth   84

8   Growth and Development of the Brain   104

9   The Interaction of Heredity and Environment in the
    Control of Growth   119

10 The Organization of the Growth Process   165

11 Standards of Normal Growth   178

12 Disorders of Growth   222

    Bibliography   240

    Index   271

To Reg Whitehouse, my friend and collaborator for twenty-nine years, whose contributions to human auxology are implicit in all these pages

Introduction

This is a book which describes the process of growth in children in terms which I hope the biologically unsophisticated reader will understand and the biologically sophisticated approve. Where complexity exists, I have said so plainly, avoiding, I hope, all facile generalizations. I have sometimes entered into details, but only when they are elegant, instructive and undisputed. Biology is like a Van Eyck painting; as you approach closer, clarity actually increases. The focus holds true down to the tiniest detail. The clothes of the street vendor seen in the far distance through the window are as clear, under the magnifying-glass, as the robes of the foreground personages talking in the room. Such perspective is not of the eye but of the mind; and I must beg the reader to use an equal facility of focus as we traverse between broad and narrow views.

Though a little biological knowledge on the reader's part is certainly helpful, I have not assumed it. Chromosomes, endocrines, proteins, codons, centiles and regressions are all explained as we go along. I have, however, assumed that readers know the names of parts of the body, including those of which they have no personal experience.

Chapter 1 presents the phenomenon we are going to study, the curve of growth of the child. It is the curve of height that is given; this is the characteristic we single out in everyday speech, for we talk of a child 'growing up'. But at the outset one must realize that growth is far from being a simple and uniform process of becoming taller or larger. As the child gets bigger there are changes in shape and in body composition and in the distribution of various tissues. In the newborn the head represents about a quarter of the total length; in the adult about one-seventh. In the newborn much of the muscle still consists of water; in the three-year-old the muscle cells are packed with contractile protein. Different tissues (such as muscles, nerves, liver) mature at different rates and the growth of a child consists of a highly complex series of changes. It is like the weaving of a cloth whose design never repeats itself. The underlying threads, each coming from its reel at its own rhythm, interact with one another continuously, in a manner always highly regulated and controlled.

The fundamental biological questions of growth relate to these processes of regulation, to the programme that controls the loom. At present very little about this is understood. In Chapter 2 we discuss cells, the fundamental units of growth, and distinguish between organs such as the

1

skin, where cells are continuously produced to replace those that continuously die, and organs such as the nerves and muscles where cells, once formed, remain throughout most of the animal's life, and no new cells can be formed once the growth period is over. There is too little certainty about the way that cell and organ growth is regulated for a long consideration of this to be included, but in Chapter 10 we return to consider the regulation of growth in the whole organism, after we have accumulated sufficient knowledge to make some sense of it. Perhaps it is worth saying to the non-biologist reader that if Chapter 2 proves difficult for him, which I hope it will not, he can take courage from the fact that all the rest is easier.

In Chapter 3, prenatal growth is described. Perhaps the most surprising fact that has emerged in the last few years is how many fertilized ova fail to result in a live baby. About 50% of all conceptuses are spontaneously aborted, nearly all because of intrinsic faults in their make-up. We have to adjust ourselves to the notion that children with certain diseases such as Down syndrome and Turner syndrome are the rare survivors from cohorts nearly all of whose members die in the first few weeks after conception.

In Chapter 4, the chromosomal and endocrine control of sex differentiation in utero and in early childhood is described, and in Chapter 5 the further differentiation that takes place at puberty. The last two decades have seen a great increase in our understanding both of the bodily changes of puberty (and especially of their great variability), and of the endocrine mechanisms which control them. The whole subject of childhood and pubertal endocrinology is at last beginning to make sense, owing chiefly to our growing ability to measure very small amounts of hormones by chemical or biological means. Further advances will now depend increasingly on studies of normal adolescents followed individually all through puberty, and that demands another and perhaps more difficult skill than the manipulation of chemical assays.

Different children experience puberty at very different ages, so that amongst a group of twelve-year-old girls or fourteen-year-old boys there will be some who have not yet started their pubertal changes and others who have practically finished them. This variability, which is almost entirely biological in origin, has important social and educational consequences. Though seen most clearly at puberty, the differences in tempo of growth – whether the course of growth is played out quickly or slowly – are present at all ages. The concept of physiological maturity, or developmental age, has thus arisen, and Chapter 6 is devoted to its description, measurement and implications.

In Chapters 7 and 8 two of the most important organ systems have been singled out: the endocrine glands because they play such an overwhelmingly important part in the control of growth and sex development, and the brain because it is of such interest and importance to human beings. Our increased understanding of endocrine events has enabled one particularly spectacular advance in medicine to be made. Some children – mostly boys – grow up to be very small, though of quite normal proportions, because they lack the growth hormone normally secreted by the pituitary gland. Normal growth is restored by injection of the growth hormone (just as injections of insulin restore the metabolic situation of diabetics).

No such spectacular advance has taken place regarding brain development. However, there is now quite a large number of research workers throughout the world devoting themselves to this technically very difficult field, and it seems safe to predict that in the next decade or so we shall have achieved real knowledge about how the complex structure of the brain is generated during growth and about the ways memories are stored. Already the work of Hubel and Wiesel and their many followers, described in Chapter 8, shows how the visual system, including the cells of the cerebral cortex, depends for its development on a precise interaction between the environment and the nerve cells. This interaction has to occur at a particular time, known as a 'sensitive period', in the animal's maturational calendar. (Sensitive periods pervade the whole of growth, as the reader will soon discover.)

Chapter 9 is a long one. It discusses the effects of the interaction of genes and environment on growth. Differences in growth between one family and another, and between one population group and another, are detailed; the effects of undernutrition are described, and of factors associated with social class, numbers of brothers and sisters, climate, psychological stress, illness and urbanization. In this field also things are rapidly becoming clearer. We are beginning to have examples of populations of all the major racial groups living under similar environmental circumstances. For example, in the USA persons of European and of African origin growing up under similar circumstances have very similar growth curves for height though not for body proportions. Under similar conditions Asians from Japan and China seem to be a little smaller and to mature a little faster. Throughout the industrialized world there has been a tendency during the last hundred years for children to become larger and reach maturity earlier, a tendency now reaching its end. Though rate of growth remains one of the most useful of all indices of public health and economic well-being in developing and heterogeneously developed countries, it must not be thought that bigger, or faster, is necessarily better. From an ecological point of view smallness has advantages; here also a change of attitude has recently been developing, with simplicity giving way to a more sophisticated search for balance.

In Chapters 11 and 12 standards for normal growth are given and some of the common pathologies of growth are briefly described. The standards provide a do-it-yourself health visitor's kit for parents interested in seeing how their children grow (comparing children's growth curves by marks made on a door specially designated for this purpose was a common pastime in large Victorian families). The proper method of measuring length or height (all too seldom followed by the health professionals) is described; how to plot the results on a chart; and how, at certain ages, to see if the child takes after the parents. Tables for predicting the probable adult height are also given. In all these standards there is very great normal variation and readers must be warned to read exactly and with caution. A famous statistician once observed that, unfortunately, there was no way of preventing half the people being below the average. Similarly, 3% of perfectly normal children are below the line on the chart called the 3rd centile, which is often taken as the arbitrary limit of normality. However, if a child is far below the standards then certainly the cause for this ought to be sought, particularly nowadays when growth-hormone deficiency, at least, can be successfully treated.

Finally, as to further reading: it seemed wiser in general not to put references in the text, where the general reader may find them distracting. I have provided instead a two-tiered bibliography for each chapter, consisting of a short list of further reading in which each paper is annotated, and a longer list of references without annotations. Where authors' names are mentioned in the text the relevant papers will be found listed in the References.

Very many colleagues and friends have helped by criticizing early drafts. I am especially grateful for the comments of Bryan Senior, Vreli Fry, Marjorie Baines and Kenneth Sass, who undertook the arduous task of reducing to a minimum the number of paragraphs incomprehensible or boring to the non-biologist. Messrs R. H. Whitehouse and Noel Cameron, and Professor Jim Garlick and his students at Cambridge, served as human biologist assessors, and Dr Michael Preece, Dr B. A. Tanner and Professor

W. A. Marshall as clinical scrutineers. The faults that remain are mine. Lastly, I wish to thank Ray Lunnon and Brian Kesterton for the pictures and diagrams, Jan Baines for assembling the material and dissembling, at times, my presence in the office, and Sara Dunford for her cheerfulness and care over innumerable typescripts.

J. M. Tanner
Institute of Child Health, London
July 1977

CHAPTER 1

The Curve of Growth

Figure 1 shows the most famous of all records of human growth. It concerns the height of a single boy, measured nearly every 6 months from birth to 18 years. This is the oldest longitudinal record in existence, and it remains, for our purpose of illustration, one of the best. It was made during the years 1759-77 by Count Philibert Guéneau de Montbeillard on his son and it was published by Buffon in a supplement to the *Histoire Naturelle*.

In Figure 1a is plotted the height attained at successive ages; in 1b the increments in height from one age to the next, expressed as the rate of growth per year. If we think of growth as a form of motion (analogous to the journey of a train) then the upper curve is one of distance travelled; the lower curve, one of velocity. The velocity, or rate of growth, naturally reflects the child's state at any particular time better than does the distance achieved, which depends largely on how much the child has grown in all the preceding years. Thus, for those substances which change in amount with age, the concentrations in blood and tissues are more likely to run parallel to the velocity than to the distance curve. Indeed, in some circumstances it is the acceleration rather than the velocity curve which best reflects physiological events.

Figure 1b shows that in general the velocity of growth decreases from birth (and actually from about the fourth month of foetal life; *see* Chapter 3), but that this decrease is interrupted shortly before the end of the growth period. At this time, from 13 to 15 years in this particular boy, there is a marked acceleration of growth, called the adolescent growth spurt. (Some writers distinguish sharply the terms 'adolescence' and 'puberty' though not all who do so agree in their distinctions. Some use puberty to refer to physical changes, adolescence to refer to psychological ones. I have used the terms interchangeably in this book.) From birth until age 4 or 5 the rate of growth in height declines rapidly, and then the decline, or deceleration, gets gradually less. A slight increase in velocity called the mid-growth, or mid-childhood, spurt occurs between 6 and 8 years in many, though perhaps not all, children, providing a second wave on the general velocity curve. It can be seen in Figure 1.

Growth is in general a very regular process. The more carefully the measurements are taken, with precautions, for example, to minimize the decrease in height that occurs during the day for postural reasons, the more regular does the succession of points in the graph become. Figure 2 shows

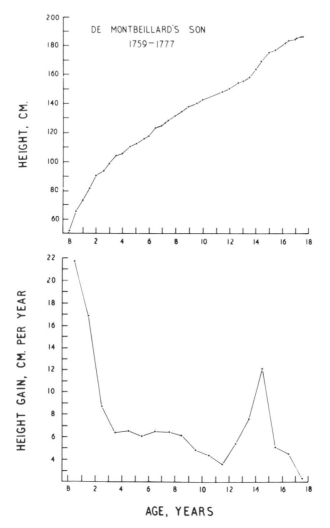

Figure 1   Growth in height of de Montbeillard's son from birth to 18 years (1759-77): (*above*) distance curve, height attained at each age; (*below*) velocity curve, increments in height from year to year
(From Tanner, 1962, drawn from data of Scammon, 1927)

the fit of a smooth mathematical curve to a series of measurements taken on a child by the same observer, my colleague R. H. Whitehouse, every 6 months from ages 3·5 to 10·0. None of the points deviates from the line by an amount more than measuring-error (which in skilled and careful hands

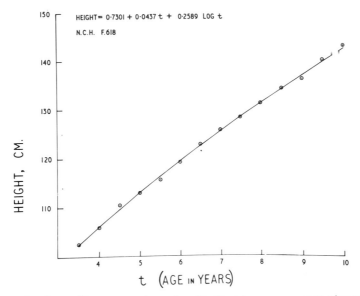

Figure 2   Curve of form $y = a + bt + c \log t$ fitted to stature measurements taken every 6 months by R. H. Whitehouse on a girl from ages 3·5 to 10·0 years (Harpenden Growth Study)
(From Israelsohn, 1960)

should seldom be more than 3 mm). This is generally true, although in some children regular seasonal variations (discussed in Chapter 9) superimpose an added 6-month rhythm about the curve. Over shorter periods of a month or so, however, there is evidence of variations in velocity. These short-term fluctuations cannot easily be picked up in total body height, but occur in measurements of the length of the lower limb from knee to heel made with I. M. Valk's knemometer, an instrument of considerable accuracy. In some short periods growth arrest or even shrinkage seems to occur, compensated by increased growth on the next or subsequent occasions. Such small fluctuations are called 'mini-spurts' by Dr Michael Hermanussen, who has recently investigated them thoroughly. Some represent the response to minor episodes of illness or mere discomfort, but others appear not to be explained this way. Over periods of three months or less, they are ironed out.

Many attempts have been made to find mathematical curves which fit, and thus summarize, human and animal growth data. Most have ended in disillusion or fantasy; disillusion because fresh data failed to conform to them, fantasy because the system eventually contained so many parameters (or 'constants') that it became impossible to interpret them biologically. What is needed is a curve or curves with relatively few constants, each capable of being interpreted in a biologically meaningful way; yet the fit to actual data must be adequate, within the limits of measuring-error. Part of the difficulty arises because the measurements usually taken are themselves biologically complex. Stature, for example, consists of leg length, trunk length and head height, all of which have rather different growth curves. Even with such relatively homogeneous measurements as forearm length or calf-muscle width, it is not clear what purely biological assumptions should be made as the basis for the form of the curve. The assumption that cells are continuously dividing leads to a different formulation from the assumption that cells are adding constant amounts of non-dividing material, or adding amounts of material at rates varying from one age period to another.

But fitting a curve to the individual values is the only way of extracting the maximum information about an individual's growth from the measurement data. This fact becomes increasingly inescapable when the effect of environmental circumstances on growth rate (e.g. illness on height growth) is investigated, or when two different measurements are being compared for the consistency of each as the child grows up. The individual's consistency can only be measured by deviations from his own growth curve. A change of rank order of two individuals in a measurement from one age to another may represent not inconsistency but consistently differing rates of change, one individual having a small velocity in the measurement and the other individual a larger one.

My colleague Professor Michael Preece, in collaboration with Dr M. Baines of Reading University, succeeded in formulating a curve which fits data on height and its components from about age two right up to maturity (provided the child has no mid-growth spurt). The curve is at first the shape shown in Figure 2, but becomes s-shaped at adolescence. In its simpler form it has only four parameters (or constants) so that all the measurements of a child from age two to maturity can be subsumed in a set of four numbers. Recently Dr Darrell Bock, of the University of Chicago, has promulgated a more complex curve, having eight parameters, which in addition fits the mid-growth spurt. All such mathematical curves have a rather fixed shape, however, and this shape may not perfectly represent the actual events. With modern computers, it is possible also to fit not mathematical curves, but curves called splines which simply smooth out – to whatever degree required – the local ups and downs of the points, just as the draughtsman draws by eye a smooth line through the points on his graph paper. This alternative approach has the advantage of flexibility, but the disadvantage of summarizing the data less well as well as less elegantly. A third possibility, recently utilized to very good effect by Dr Theo Gasser of Mannheim, is to use an advanced form of smoothing by weighted moving averages. Both splines and moving averages have been made practical by the introduction of more powerful and faster computers. Locating the various points in the curve that are of special interest – age at peak height velocity in the adolescent spurt for example – can be simply read off by the computer programme, and in this rather new situation the importance of having a curve with a small number of parameters suddenly seems much less.

Types of Growth Data

These curves are designed to be fitted to data on single individuals. Curves of the yearly averages derived from different children, each measured once only, in general have a different shape. Thus the distinction between the two sorts of investigation is very important. The method of study using the same child at each age is called *longitudinal*; that using different children at each age is called *cross-sectional*. In a cross-sectional study each child is measured only once and all the children at age 10, for example, are different from those at age 9. A longitudinal study may extend over any number of years; there are short term longitudinal studies extending from ages 3 to 5, for instance, and full birth-to-maturity longitudinal studies in which the children may be examined once, twice, or even more times every year from birth until 20 or over.

In practice it is always impossible to measure exactly the same group of children every year for a prolonged period; inevitably some children leave the study, and others, if that is desired, join it. A study in which this happens is called a *mixed longitudinal* study, and special statistical techniques are needed to get the maximum information out of its data. One particular type of mixed study is that in which a number of relatively short-term longitudinal groups are interlocked; for exampe, we may simultaneously study four groups, one followed from 0 to 6 years, the second from 5 to 11, the third 10 to 16, and the fourth 15 to 20. Thus the whole age-range is covered for estimates of mean yearly velocity in the research time of five years. However, problems arise at the 'joins' unless the sampling has been remarkably good.

Both cross-sectional and longitudinal studies have their uses, but they do not give the same information and cannot be dealt with in the same way.

Cross-sectional surveys are obviously cheaper and more quickly done, and can include far larger numbers of children. They tell us a good deal about the curve of growth of height or weight for a given age (the 'distance' curve). It is essential to have cross-sectional surveys as part-basis for constructing standards for height and weight and other measurements in a given community, and periodic surveys are valuable in assessing the nutritional progress of a country or of particular socio-economic groups, or the health of the child population as a whole. But cross-sectional surveys have one great drawback; they can never reveal individual differences in rate or velocity of growth or in the timing of particular phases such as the adolescent growth spurt. It is these individual differences in growth velocity which chiefly throw light on the genetical control of growth and on the correlation of growth with psychological development, educational achievement and social behaviour. Longitudinal studies are laborious and time-consuming; they demand great perseverance on the part of those who make them and those who take part in them; and they demand very high technical standards, since, in the calculation of a growth increment from one occasion to the next, two errors of measurement occur. They demand also a sequential approach to problems, with all past data fully computerized and available for analysis in relation to a specific question at any time. The evidence of the past suggests that unless accompanied by cross-sectional surveys and animal experimentation, as they were, for example, at the Fels Research Institute in Yellow Springs, Ohio and at the Department of Growth and Development of the Institute of Child Health of London University, they can sink over the years into sterile deserts of number-collecting. But they are indispensable.

Cross-sectional data can be in some important respects misleading. Figure 3 (p.12) illustrates the effect on 'average' figures produced by the individual differences in the age at which the adolescent spurt begins. Figure 3a shows a series of individual velocity curves from 10 to 18 years, each individual starting his spurt at a different age. The average of these curves, obtained simply by treating the values cross-sectionally and adding them up at ages 10, 11, 12, etc. and dividing by 5, is shown by the dashed line. The line in no way characterizes the 'average' velocity curve; on the contrary, it is a travesty of it. It smoothes out the adolescent spurt, spreading it along the time axis. It does not take account of the 'phase-differences' between the individual curves. Figure 3b shows the same individual curves, but arranged so that their peak velocities coincide; the average curve then characterizes the group in a proper manner. In passing from Figure 3a to 3b the time scale has been altered so that in 3b the curves

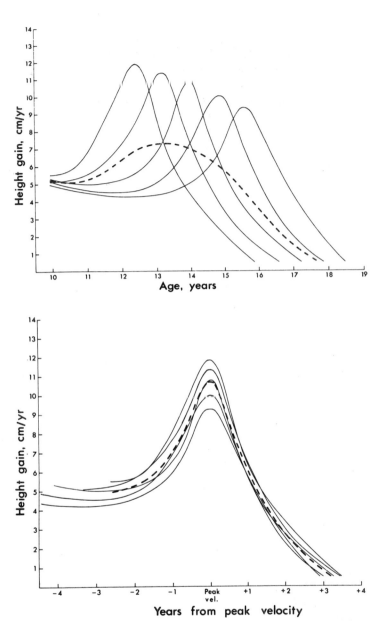

Figure 3   The relation between individual and mean velocities during the adolescent spurt: (*above*) the individual height velocity curves of five boys of the Harpenden Growth Study (solid lines) with the mean curve (dashed) constructed by averaging their values at each age; (*below*) the same curves all plotted according to their peak height velocity (From Tanner, Whitehouse and Takaishi, 1966)

are plotted not against chronological age but against a measure which arranges the children according to how far they have travelled along their course of development; in other words, they are arranged according to their true developmental or physiological growth-status. This is nearly always the appropriate method in analysing longitudinal data, especially at adolescence.

Averages computed from cross-sectional data, however, inevitably produce velocity curves of this flattened, distorted type; and, equally, distance curves show the distortion by not rising sufficiently rapidly at adolescence.

Until recently all the published height and weight standards used in hospitals and schools incorporated this distortion. However, it is possible to construct a curve which represents the actual growth of a typical individual, by taking the shape of the curve from individual longitudinal data, and the absolute values for the beginning and end from large cross-sectional surveys. Figures 4 and 5 show height-attained and height

Figure 4   Typical-individual height-attained curves for boys and girls (supine length to the age of 2; integrated curves of Figure 5)
(From Tanner, Whitehouse and Takaishi, 1966)

Figure 5 Typical-individual velocity curves for supine length or height in boys and girls.
These curves represent the velocity of the typical boy and girl at any given instant
(Modified from Tanner, Whitehouse and Takaishi, 1966)

velocity curves for the 'typical' boy and girl in Britain in 1965, determined
in this way. By 'typical' is meant a boy or girl who has the mean birth
length, grows always at the mean velocity, has the peak of the adolescent
growth spurt at the mean age, and, finally, reaches the mean adult height
at the mean age of cessation of growth. There is, of course, a certain danger
in showing such a smooth, average curve, for measurements on a single
individual are naturally less regular, however expert the measurer. Almost
no individual exactly follows the curves of Figures 4 and 5; but most
individuals have growth curves of this shape. Individuals with mid-growth
spurts have a slight hump in their curves around age seven, but these have
been ignored in this general presentation. Standards for height, constructed
around these curves, are given in Chapter 11.

Boys' and Girls' Height Curves

Figures 4 and 5 show the height curves from birth to maturity. Up to age 2 the child is measured lying down on his back. One examiner holds his head in contact with a fixed board, and a second person stretches him out to his maximum length and then brings a moving board into contact with his heels (*see* Figure 63, p.186). This measurement is called supine length, and averages about 1 cm more than the measurement of standing height taken on the same child, even when, as in the best techniques, the child is urged to stretch upwards to full height and is aided in doing so by a measurer applying gentle upward pressure to the bony prominences behind the ears (Figure 62, p.183). This difference in method of measurement causes the break in the line in Figure 4 at age 2.

Figure 4 shows the typical girl as slightly shorter than the typical boy at all ages until adolescence. She becomes taller shortly after 11·0 years because her adolescent spurt takes place two years earlier than the boy's. At age 14·0 she is surpassed again in height by the typical boy, whose adolescent spurt has now started, whereas hers is nearly finished. In the same way, the typical girl weighs a little less than the boy at birth, equals him at age 8·0, becomes heavier at age 9·0 or 10·0 and remains so till about age 14·5.

The velocity curves given in Figure 5 show these processes more clearly. At birth the typical boy is growing slightly faster than the typical girl but the velocities become equal at about 7 months and then the girl grows faster until about age 4·0. From then till adolescence no difference in velocity can be detected. The sex difference is best thought of, perhaps, in terms of acceleration, the boy decelerating harder than the girl over the first four years. The typical girl begins her adolescent height spurt at about 10·5 and reaches peak height velocity at approximately 12·0 in the UK and about six months earlier in the USA. The boy begins his spurt and reaches his peak just two years later. The boys' peak is higher than the girls', averaging in our data 10·3 centimetres a year compared with the girls' 9·0 cm/yr (as 'instantaneous' peaks, i.e. peaks obtained by fitting smooth curves to the observations; the velocities over the whole year which includes the peak moment are naturally less, averaging 9·5 cm/yr for boys and 8·1 cm/yr for girls).

Girls are always in advance of boys (i.e. closer to their final mature status), even at birth; this very important sex difference is considered in more detail in Chapters 5 and 6.

Growth Curves of Different Tissues and Different Parts of the Body

Most body measurements follow approximately the growth curves described for height. The great majority of skeletal and muscular dimensions grow in this manner, and so also do the internal organs such as liver, spleen and kidney. But some exceptions exist, most notably the brain and skull, the reproductive organs, the lymphoid tissues of the tonsils, adenoids and intestines, and the subcutaneous fat.

In Figure 6 these differences are shown, using the size attained by various tissues as a percentage of the birth-to-maturity increment. Height follows the 'general' curve. The reproductive organs, internal and external, have a slow pre-pubescent growth, followed by a very large adolescent spurt.

Figure 6   Growth curves of different parts and tissues of the body, showing the four chief types. All the curves are of size attained, and plotted as percentage of total gain from birth to 20 years, so that size at age 20 is 100 on the vertical scale.
Lymphoid type: thymus, lymph nodes, intestinal lymph masses
Brain and head type: brain and its parts, dura, spinal cord, optic apparatus, cranal dimensions
General type: body as a whole, external dimensions (except head), respiratory and digestive organs, kidneys, aortic and pulmonary trunks, musculature, blood volume
Reproductive type: testis, ovary, epididymis, prostate, seminal vesicles, fallopian tubes
(From Tanner, 1962, redrawn from Scammon, 1930)

AGE, YEARS

The brain, together with the skull covering it and the eyes and ears, develops earlier than any other part of the body and thus has a characteristic postnatal curve. (Brain growth is further discussed in Chapter 8). If the brain has any adolescent spurt at all, it is a very small one. A small but definite spurt occurs in head length and breadth, but most or all of this is due to thickening of the skull bones and the scalp, and development of the air sinuses. The dimensions of the face follow a path somewhat closer to the general curve. There is a considerable adolescent spurt, especially in the mandible, resulting in the jaw's becoming longer and more projecting, the profile straighter, and the chin more pointed. But, as always in growth, there are considerable individual differences, to the degree that a few children have no detectable spurt at all in some face measurements.

The eye probably has a slight adolescent spurt, although present data are not sufficiently accurate to make it certain. Very likely it is this that is responsible for the increase in frequency of short-sightedness which occurs at the time of puberty. Although myopia increases continuously from at least age 6 to maturity, a particularly rapid rate of change occurs at about 11 to 12 in girls and 13 to 14 in boys, as would be expected if there was a rather greater spurt in the axial dimension of the eye than in its vertical dimension.

The lymphoid tissue has quite a different growth curve from the rest of the body. It reaches its maximum amount before adolescence and then, probably under the direct influence of sex hormones, declines to its adult value.

The subcutaneous fat layer also has a curve of its own, of a slightly complicated sort. Its thickness can be measured either by X-rays or ultrasound or, more simply, at certain sites in the body, by picking up a fold of skin and fat between the thumb and forefinger and measuring its thickness with a special, constant-pressure caliper. Figure 7 shows the distance curves of skinfolds taken half-way down the back of the arm (triceps) and at the back of the chest, just below the shoulder blade (subscapular). Subcutaneous fat begins to be laid down in the foetus at about 34 weeks postmenstrual age (*see* p. 35), and increases continuously, to reach a peak at about 9 months after birth. (This is in the average child; the peak may be reached as early as 6 months or as late as 12 or 15.) After 9 months, the skinfolds decrease until age 6 to 8 when they begin to rise again. Girls have a little more total fat than boys at birth, and the difference becomes gradually more marked during childhood. From 8 years on, the curves for girls and boys diverge more radically, as do the curves

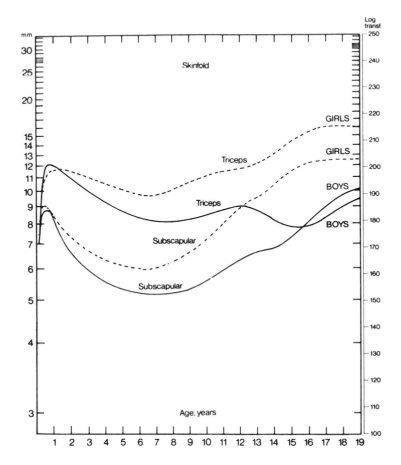

Figure 7   Distance curve of subcutaneous tissue measured by Harpenden skinfold calipers over triceps (back of upper arm) and under scapula (shoulder blade). Scale is mm on the left and logarithmic transformation units on the right-hand side. British children, 50th centiles
(From Tanner and Whitehouse, 1975)

for limb and body fat. At adolescence the limb fat in boys on average decreases (*see* triceps, Figure 7); the body fat (subscapular) shows a temporary slowing down of gain, but no loss. In girls there is a slight halting of the limb-fat gain at adolescence, but no loss; the trunk-fat shows only a steady rise until adulthood. The range of skinfold thickenesses at each age from birth onwards in English children is given in Tanner and Whitehouse (1975), and in United States schoolchildren in Johnston *et al.*, (1974).

Postadolescent Growth

Growth, even of the skeleton, does not entirely cease at the end of the adolescent period. The limb bones stop increasing in length, but the vertebral column continues to grow until about age 30, by apposition of bone to the tops and bottoms of the vertebral bodies. Thus height increases by a small amount, on average 3 to 5 millimetres. From about 30 to 45 years height remains stationary, and then it begins to decline. Head length and breadth and facial diameters increase slightly throughout life; and so do the widths of the bones in the leg and in the hand, in both sexes. For practical purposes, however, it is useful to have an age at which we may say that growth in stature virtually ceases, i.e. after which only some 2% is added. At present in North America and north-west Europe the average boy stops growing, in this sense, at 17·5 years and the average girl at 15·5 years. There is a normal range of variation amongst individuals, amounting to about two years, on either side of these averages.

Individual Variation in Growth

There is very wide variation amongst children both in height at any age and in the velocity of height growth from one age to the next. The extent of this variation is described in some detail in Chapter 11, where charts enabling readers to place themselves and their children in relation to others are given. It is necessary here, however, to give a brief description of the way in which individual variation is measured, since we shall need to have some of the terminology in mind during many of the succeeding chapters.

Figure 8 shows the standards for height at successive ages for boys in the UK. Each of the lines represents a *centile* (or percentile). Their meaning is as follows. Imagine 100 boys all aged 9·0 years. Measure them and line them up precisely in order of increasing height. Then place a line marked 3 so that 3 out of the 100 lie just below it. Place further lines such that 10 boys lie below the 10th centile line, 25 below the 25th centile and just half of the boys below the 50th centile line. At the top end 97 boys lie below the 97th centile (i.e. only 3 boys lie above it). Notice that this system can be applied whatever the shape of the curve traced out by the heads of those 100 boys lined up in order of increasing height. If this line of heads rose at first gently, then much more suddenly at the higher values, the centile approach would still be valid.

Readers who are familiar with statistics (and those to whom statistics is a closed book will miss nothing vital by skipping the remainder of this paragraph) will know that the 50th centile is the same as the mean in the

Figure 8   Centile standards for boy's height, birth to 19 years: British data
(From Tanner and Whitehouse, 1976)

Normal or Gaussian distribution. Height is in fact distributed in this way
and it is for this reason that conventionally the outside centiles are taken
as the 3rd and 97th, which are rounded forms of the 2·5th and 97·5th,
corresponding respectively to plus and minus 2 standard deviations

(analogous to the 5% probability level in a two-sided significance test). But weight is not distributed Normally at most ages, and nor are skinfolds. For these distributions the centile positions are still valid while standard deviation scores are not. The accuracy with which the centiles are located, however, depends on the distribution, the standard error of the centile being greater the more the distribution departs from Normal. Conventionally we say that a child below the 3rd centile should be regarded as possibly pathological. It must be stressed that this is a convention and nothing more; there is no magic in the number 3, at least in this context. If the next step in determining whether a child was abnormal was something drastic like an exploratory operation, then clearly we would demand a much lower level than the 3rd or even the 1st centile, to avoid operating on children who were normal. If, on the other hand, it only meant giving extra morning milk and biscuits, we might choose the 10th centile to make sure we got all the undernourished within our treatment net. If we set up very stringent standards, like the 1st centile, we run the risk of leaving some truly pathological children in amongst the healthy ones. We shall see further how the theory of standards works in practice in Chapter 11.

The Human Growth Curve as a Primate Characteristic

The characteristic shape of the human growth curve is shared only by apes and monkeys, the other members of the order *Primates*. It is apparently a distinctive primate characteristic, for neither rodents nor cattle have curves resembling it.

In Figures 9 and 10 are shown the velocity curves for body weight of mice and chimpanzees (weight has to be used, since so few data on length exist). In the mouse, as in the rat, there is little interval between weaning and puberty, and no visible adolescent spurt because there is no period of low velocity between birth and maturity. In terms of the maturation of their organs, mouse and rat are born earlier in development than man. The peak velocity of the mouse's weight curve occurs at a time that corresponds closely, by this organ-maturation calendar, with birth in man, which is when the first peak of man's weight velocity curve occurs. In the chimpanzee (Figure 10), on the other hand, the curve resembles entirely that of man. The first peak velocity of weight is shortly before, or at, birth, but this is followed by a gradual decrease of velocity during the long interval between weaning and puberty. The rhesus monkey has a similar curve, though with less time intervening between weaning and puberty.

Figure 9   Weight velocity curve for the mouse: curve 1, sexes combined, cross-sectional; curves 2 and 3, males (18) and females (18), pure longitudinal, MacArthur large strain (From Tanner, 1962, p. 228, where sources of data are detailed)

Figure 10   Weight velocity curve for the chimpanzee. Actual increments, from Grether and Yerkes (From Tanner, 1962, p. 234, where sources of data are detailed)

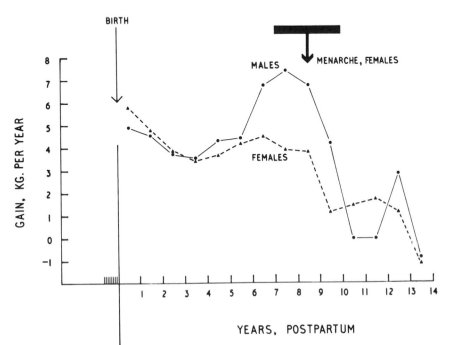

The prolongation of the time between weaning and puberty appears to be an evolutionary step taken by the primates, reaching its most pronounced development in Man. The increased time necessary for the maturing of the primate brain has been sandwiched into this period. At least some of the evolutionary reasons are not hard to find. It is probably advantageous for learning, especially learning co-operation in group and family social life, to take place while the individual remains relatively docile and before he comes into sexual competition with older individuals.

CHAPTER 2

Cells and the Growth of Tissues

All biological tissues are made up of cells. Even the skeleton is scarcely an exception, for the mineral deposits which are the characteristic of bone are laid down around an intricate network of bone cells, whose activity continuously adds and subtracts, remodels and repairs; if the bone cells die the skeleton soon becomes unserviceable. There are very many different sorts of cell in the body, but all have the same fundamental architecture and contain the same sort of machinery. Metaphors follow the economic and cultural activity of the age, and the modern picture of a cell is a factory with numerous production lines, each turned on or turned off by messengers arriving with instructions from the outside world.

Muscle and nerve cells fit this description only with some qualification. Most cells export their products, for use in particular bits of the general body economy; but muscle cells manufacture only their own form. Muscle is like a stack of playing-cards, with red and black cards stacked alternately. All the red cards project a little to one side, all the black cards a little to the other. Muscle movement corresponds to the cards sliding inwards again to make the edges of the pack even. It takes energy to pull the cards in in this way and it is this energy that is the product of the muscle-cell production lines. As for nerve cells, they do indeed have manufacturing lines producing substances which are extruded, but they also have the specialized function of electrical impulse transmission and this also requires energy.

Regenerating and Non-Regenerating Tissues

Muscle and nerve cells are peculiar also in that neither tissue is able to manufacture new cells once the growth period is over. When a cell is formed it lasts for most or all of the animal's lifetime. In many tissues, on the other hand, cells are continuously dying and being replaced by new cells. Examples are skin cells, blood cells and the cells lining the gut. Most such tissues have a special germinative zone where new cells are continuously manufactured.

In a third class of tissue, cells are relatively long-lived and stable but new cells can be made when the tissue is damaged, or when the work load on the tissue is greatly increased. Most glands (including the endocrine glands described in Chapter 7), and parts of the liver and kidney come into this category. In these tissues there is no specialized germinative zone;

cells throughout the whole tissue, it seems, are capable of dividing when necessary, at any time throughout the animal's life. Thus liver has a marked power of regeneration; in the rat, a large proportion of it can be removed every month for a year or more, and every month the liver will regenerate itself. (Evidently the Greeks knew this, for when they created the myth of eternal punishment for Prometheus, it was his liver which the vulture pecked out each night. His pancreas would have done as well, but only that portion concerned with digestion and not the insulin-providing islets of Langerhans (p.100); if the vulture had devoured them Prometheus would have died of diabetes.)

Muscle, however, has very little such power of regeneration, and nerve cells no power of regeneration at all. Nerve cells wear out (to use a vague term, indicating, perhaps, the occurence of something analogous to metal fatigue), die and are lost. But they are not replaced.

The Dynamic State of the Body Constituents

All cells, however, have to renew their constituents, including even the constituents that make up the structural skeleton of the cell. There is a continual flow of substance into and out of the human body so that most molecules entering in the food stay only a relatively short time before being excreted. Nitrogen, for example, is mostly present in protein molecules, and an average adult man has a protein turnover of around 230 grammes per day; that is to say, he loses this amount, and replaces it with an equivalent amount of newly made protein. (Only a proportion of this is newly *ingested*, the rest being built up again from the remnants of broken-down protein: on a high protein diet the newly ingested proportion is about a half, while on a low protein diet it falls to a third or less; the absolute rate of turnover is little affected by diet.) Most of this turnover occurs in muscle, so that some 2 to 3% of muscle protein is replaced every day. In children the turnover is greater, especially during infancy, when the rate is about three times that of the adult. Different substances turn over at different rates; while the average nitrogen molecule remains in the body only a week or two, a molecule of calcium, the major constituent of bone, remains in the body a matter of months. A few substances, for example the cholesterol laid down in growing nerve sheaths, seem never to leave the body at all, but this is exceptional.

Thus the body is in a constant state of flux; we do not consist of certain particular molecules, but of a pattern imposed on continuously changing molecules. This dynamic state enables us to adapt to a continually changing environment, which presents now an excess of one type of food, now an excess of another; which demands different levels of activity at different times and which is apt to damage the organism. But we pay in terms of the energy we must take in to keep the turnover running. Protein turnover accounts for much of our basal energy expenditure, i.e. the energy we produce as heat when totally at rest. Enough food must be taken in to provide this energy, or the organism begins to break up.

Cell Structure

The modern picture of a cell is a product of the electron microscope, which has revealed structural details quite unknown before. Figure 11 shows a diagrammatic reconstruction. The cell is surrounded by a membrane. Attached to the membrane are large numbers of receptor molecules, each specific for a particular chemical substance. Thus when a molecule of growth hormone (*see* p.89), for instance, passes by a particular type of liver cell, it is seized by its receptor on the liver cell membrane. Its capture starts a series of events in the cell which results in the turning on of the production line for a hormone called somatomedin (p.92) which is then secreted into the blood stream. Thus the nature of the cell membrane is exceedingly important in the specialized function of the cell.

Inside the cell the most prominent structure is the nucleus. With few

Figure 11   Diagram of simple cell, to show important structures and assembly of protein molecules on ribosomes under DNA/RNA control

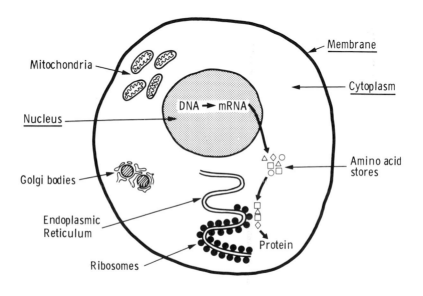

exceptions (*see* muscle, p.29) cells have only one nucleus. Within the nucleus lie the chromosomes, invisible most of the time, because they are very long thin strands except when a cell is dividing. When a cell divides the chromosomes coil up like springs into short, thick and identifiable strands (*see* Figure 20, p.53). These strands split longitudinally at cell division in a tissue like the liver, so that each daughter cell has the same chromosomes as the mother cell. Since the chromosomes control the fundamental nature and working of the cell this ensures that like cell begets like cells (except at one particular cell division, occuring only in germ cells, that need not concern us here). The chromosomes carry the genes, traditionally visualised as cylindrical beads strung, at irregular intervals, on the chromosomal threads.

Each gene consists of a length of DNA, deoxyribonucleic acid, a molecule resembling a rope ladder twisted into spiral form. The rungs of the ladder are of two different types. One is formed by a bonding of the substances thymidine and adenine, each substance sticking inwards from one side of the ladder. The other is similarly formed of two other substances, guanine and cytosine. Each rung may be arranged with one or the other end at the left- or right-hand side of the ladder; thus there are T-A, A-T, G-C, and C-G rungs. When a strip of DNA multiplies, the bondings break and the molecule, as it were, unzips. Half of the spiral ladder is free, with half-rungs hanging loose. A new piece is then manufactured exactly to correspond; only then can the molecule be zipped up again.

The rungs measure approximately one-millionth of a millimetre from one end to the other and are placed up the ladder at intervals of about a third of this distance. The ladder spirals gently, so it takes ten rungs for one complete turn.

A gene consists of a length of DNA containing typically about a thousand rungs. Since the rungs may be arranged in any sequence, the number of possible arrangements of a thousand rungs is practically infinite, and each gene has its own unique structure. It is the precise order of the rungs which constitutes the fundamental genetic instruction. Mutation is a change from one sort of rung to another. In many cases the number of mutations necessary to bring about the differences in structure between for example, bovine and human growth hormone (*see* Chapter 8) can be estimated fairly reliably. The affinities between different species and the speed of evolution are beginning to be thought of in these molecular terms.

It is important to realize that what is inherited is DNA, not blue eyes nor tallness nor even blood group A. The DNA of the chromosome acts as a template for the building of a similar molecule, RNA, ribosenucleic acid.

After alterations eliminating inactive sections, this leaves the nucleus and enters the machine-shop of the cytoplasm (*see* Figure 11). Here it activates that particular production line which assembles proteins from the amino-acids stored in some cellular pigeon hole. Proteins consist of strings of amino-acids, often folded into extraordinary and beautiful shapes. There are only twenty amino-acids and millions of proteins, so the identity of a protein depends largely on the order and the number of its constituent amino-acids. It is this that the RNA controls. Each group of three consecutive rungs of the RNA ladder corresponds to, or 'specifies', a particular amino-acid (apart from groups of rungs which carry 'start-making' or 'stop-making' instructions). For this reason the system is often referred to as the 'triplet code' and each three-rung group is called a 'codon'. In this way the order in which the amino-acids are joined one to another, and hence the nature of the protein produced, is dictated by the RNA controller, and thus reflects the original instructions of the DNA. Thus a one-to-one relation exists, barring accidents, between the gene and the amino-acid sequence constituting the protein. Normally, therefore, we can say that blood group antigen A, which is such a protein, is directly gene-controlled, that is, inherited. Even in such cases, however, the working protein is often not quite the protein run off under the RNA controller, but a more complicated molecule, with pieces added or subtracted by other sorts of cell machinery. Thus interactions between the gene product and the environment begin at the level of the cell machinery itself, and continue through the level of cell interacting with cell to the level of whole organism interacting with the outside environment.

The actual production lines appear to be the endoplasmic reticulum, a membrane-like structure along which are strung the ribosomes, analogous to work bays, at each of which a specific task of assembly is carried out.

Cell Growth and Organ Growth

An organ may enlarge either because the number of its constituent cells increases, or because the content of each of its cells increases, or because the amount of inert intercellular substance increases. In all non-regenerating tissues, growth goes through three phases (Figure 12). First, cell division and increase in cell numbers occurs, without any change in the amount of material in each cell. At this stage the tissue consists almost entirely of cell nuclei, with much intercellular fluid. The cells have little cytoplasm. In the second phase, the rate of cell division decreases; however, proteins continue to be synthesized at much the same rate as before. They now enter

Figure 12 Phases of cell and organ growth: phase 1, multiplication of all cells without addition of cytoplasm; phase 2, multiplication of some cells and addition of cytoplasm; phase 3, no multiplication of cells, further addition of cytoplasm

the cytoplasm and the cells become larger. The intercellular substance decreases correspondingly. New cells continue to be made, but at a slower rate than before. In the third phase, all division ceases and cells grow only in size. In all non-regenerating tissues, cell division (or mitosis) ceases some time before the cells stop increasing in size. The age at which cells cease dividing varies according to the tissue and organ considered.

Nerve   The first cells to stop dividing in the foetus are probably the neurons (or nerve cells proper) of the central nervous system, whose division is finished by about 18 postmenstrual weeks in the cerebral cortex (*see* Chapter 8, p.104). The supporting tissue of the nerve cells, however – the neuroglia, which provide a scaffolding and serve the nerve cells in some such way, presumably, as bone cells serve bone – continue to multiply until one or two years after birth (*see* p.107).

Muscle   Muscles are exceptional in that muscle fibres, the basic units, are formed by fusion of several cells during development, a situation unique amongst developing tissues. In consequence, each fibre has more than one nucleus. The number of muscle fibres is fixed early in life, like the number of nerve cells. The size of the fibres increases, however, with the manufacture and incorporation of contractile protein. The number of nuclei seems also to increase, at least in man, keeping pace, more or less, with the increase in fibre size. At puberty in boys, both fibre size and number of nuclei increase considerably. However, these nuclei do not result from division of the exisiting muscle-fibre nuclei, which, once formed, like nerve-cell nuclei are incapable of dividing. Surrounding but outside the muscle fibres are numbers of nuclei which represent leftovers of the cells which formed the fibres in embryonic life. These cell nuclei are incorporated into the enlarging fibre, apparently differentiate into mature muscle-fibre nucleis, and then lose the ability to divide or differentiate further (*see* Chapter 10). Thus muscle carries a reserve supply of cells which even in adult life can provide a small potential for regeneration and repair. In this it resembles adipose rather than nerve tissue. It is not known for certain whether these reserve (or 'satellite') cells can multiply their number in childhood or in adult life in response to growth demand or injury, as can the cells of the cambium layer of the periosteum which covers bones (*see* p.32).

It is also rather uncertain whether all the nuclei newly appearing in muscle at adolescence are really in the muscle fibres themselves; their presence has only been inferred from chemical analysis, which shows an increase of DNA. DNA is present solely in nuclei; and all nuclei have approximately the same amount. Thus the amount of DNA tells one how many nuclei are present in a specimen. But it does not distinguish the nucleus of one type of cell from the nucleus of another, and some of the increase may represent connective tissue cells. Chemical analysis has to be viewed with caution; though its use enables many more experiments to be done in a given time, control of the results by electron and optical microscopy is still all-important.

Fat Fat cells can be counted in biopsies of the buttock fat obtained very simply by inserting a small needle and sucking fat up into it by means of an ordinary syringe. The fat cell is peculiar in that its nucleus is quite small, while its cytoplasm may enlarge to enormous bulk, carrying just one droplet of stored fat. The average size of fat cells does not appear to increase during pre-pubertal childhood, but the number of fat cells that can be counted increases steadily up to about the time of puberty. However, most present-day techniques for counting the number of cells depend on the cell having reached a certain size, i.e. being filled with a certain amount of fat. Thus it may be that fat is like muscle, with embryonic nuclei left over in the tissue which cannot at present be identified or counted. As growth progresses, an increasing proportion of these may fill up with fat and become identifiable. If so, adipose tissue would be similar to muscle and nerve in that the period during which its nuclei could divide would terminate early in development, perhaps even before birth. This seems likely, since fat cells once formed do not appear to die, and in this they resemble muscle and nerve rather than regenerating tissues such as skin and glands.

Interference With the Cellular Clock
Professor Myron Winick, Dr Jo-Anne Brasel and their colleagues at Columbia University have shown in a long series of studies that severe undernutrition may stop cells dividing if it occurs at a time when cell division is normally going on. When the undernutrition is terminated cell

division may proceed, but only at the rate appropriate to the age of the animal. This rate may be zero if the undernutrition has gone on long enough. Thus undernutrition may cause a permanent deficit in the number of cells. On the other hand, undernutrition occuring during the phase when cells are only increasing in size seldom has permanent effects. It seems that in this phase animals can simply stop the mechanism of protein storage in cells and then switch it on again, usually at a faster tempo than normal, when circumstances improve. We do not understand why the mechanism of cell division cannot be frozen in the same way (or at any rate to the same degree). The clock controlling it seems inexorably to continue whether or not materials are available to carry out the programme.

Whether overnutrition can cause an increase in the number of fat cells is not known. It has been said that overnutrition applied at the time when the cells are still being formed may indeed do so. However, as we have seen, it is not clear whether new fat cells are continuously formed during childhood, or whether existing embryonic cells are simply being filled up with fat to a level at which they can be seen. The issue is important, since it might be that the overfeeding of babies, prevalent in some parts of the world, induces an excessive number of fat cells as well as an excessive amount of fat in the cells. The distinction is crucial; hard though it may be to reduce the fat in an existing cell, it is at least possible, while actually to lose a fat cell probably is not. Thus obesity due to an excessive number of cells would be more intractable than obesity due to excessive amounts of fat in a normal number of cells.

Increase of muscle bulk may be regarded with more favour than increase in fat. Muscular training by heavy exercise does not cause an increase in the number of muscle fibres, but only in their size. However, training is applied (in man) at ages when cell division is believed to have ceased. It is an open question whether hormones have an influence on the number of muscle fibres in the foetus or even in the immediately postnatal child. If they do, it would probably be the androgenic or male-determining hormones such as testosterone which would be concerned (*see* Chapter 7), and it is possible that the effect might be permanent. Little research has been done on this aspect of athletics.

Growth of Bones
The manner in which bones grow merits a slightly longer description, both because it is bone lengths that we so often measure in our studies of growth, and because bone is an exceptional organ from the point of view of cellular growth.

A limb bone, such as the radius in the forearm, is laid down first as a cartilage model. At the centre of this model the cartilage cells break down and bone appears. This area is called the primary centre of ossification; in most bones it appears quite early in foetal life. Later, starting shortly before birth, secondary centres of ossification appear, mostly at the two ends of the long bones. They also represent conversions of pre-existing cartilage into bone, and are called epiphyses. New epiphyses appear in various parts of the skeleton right up to puberty.

The situation is then as shown in the diagram of Figure 13 and the first radiograph of Figure 14. The latter represents the distal or wrist end of the radius together with its epiphysis. The cartilage between these two bone areas is translucent to the X-rays and hence looks black in the radiograph. This area is called the growth-plate.

In the part of the growth-plate immediately under the epiphysis is a layer of cells called stem cells or chondroblasts. These cells are the origin of the columns of cells which stretch downward through the proliferative cell layer to the hypertrophic cell layer (*see* Figure 13b). Each column of cells is derived from a single stem cell and is thus what is called a clone. The stem cells divide relatively infrequently, but when one does its

Figure 13(a)   Diagram of limb bones with upper and lower epiphyses

Figure 13(b)   Magnification of epiphysis-shaft junction to show zones of cells. New cells are formed in proliferation zone and pass to the hypertrophic zone to add to the bone accumulating on top of the bone shaft

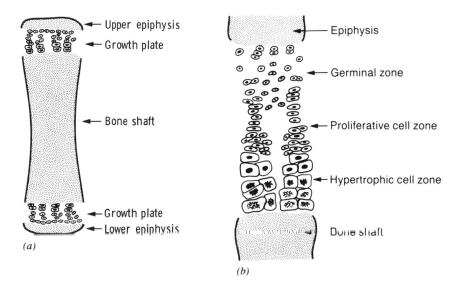

Upper epiphysis
Growth plate

Bone shaft

Growth plate
Lower epiphysis

(a)

Epiphysis

Germinal zone

Proliferative cell zone

Hypertrophic cell zone

Bone shaft

(b)

daughter cell differentiates into the next cell in the series, the proliferative cell, and at this stage in its life cycle divides more often. The division and differentiation of the stem cells is increased by growth hormone (*see* Chapter 7) whereas the division in the proliferative phase is stimulted by somatomedin C, also called IGF-1 (for insulin-like growth factor 1) which the cells themselves produce under the stimulus of growth hormone. It is thought that a stem cell produces a clone of some 60 or 100 cells; each of these moves to successively lower positions in the column because the last cell of the column is continually being absorbed into the bone shaft, and thus enlarging it. After the proliferative cell has passed about half way to its destination and death it switches abruptly to a massive hypertrophy, enlarging to a diameter some four times greater than previously. It then secretes substances which induce mineralization of its surroundings and after the cells ahead of it have been incorporated in the bone, meets its own fate.

This journey takes a cartilage cell about 20 days in Man – at least in the only epiphysis we know about, the one at the lower end of the femur, above the knee-joint (*see* Kember and Sissons, 1976). When all the progeny of a single stem cell have completed their journeys it is the clone from the next generation of stem cells which takes up the baton. It is at present believed that differences of rate of growth with age are due to differences in the rates of production of new cells in the proliferative zone. That one epiphysis grows faster than another, at the same age, is thought to be due to differences in the actual lengths of the proliferative zones in the different areas. The basis of differences in skeletal shape between one person and another or one species and another would reside, in this view, on the factors (whatever they are) controlling the number of cells in the cartilage clone.

In a radiograph of a rapidly growing bone (Figure 14a) the zone immediately next to the main shaft can be clearly seen, as a specially dense (i.e. white) band, because it is there that calcium, opaque to the X-rays, is being laid down at maximal concentration.

As growth slows down the growth-plate gets thinner (Figure 14b). Finally bone from the main shaft breaks through it, reaches the bone of the epiphysis, and soon eliminates the plate altogether (Figure 14c). When the plate has gone, no more growth in limb bones is possible. The ephiphyses are said to be closed, or 'fused' (i.e. joined to the main shaft).

In some animals such as rats the limb bone growth-plates may never completely disappear, and growth may continue very slowly even into old age. In man it is the sex hormones which bring about the final elim-

(a) *(b)*

(c)

Figure 14(a) Radiograph showing epiphysis of radius at wrist; black area represents cartilage growth plate

Figure 14(b) Growth plate narrowed and zone of newly forming bone at end of shaft prominent as dense white band

Figure 14(c) Growth plate practically eliminated, 'epiphysis fused with shaft'

ination of the plates. Children who lack sex hormones remain with open epiphyses (and hence a pre-pubertal bone age – *see* Chapter 7). However, the capacity for such children to grow is not very great; growth is not stopped mechanically by closure of the epiphyses as used to be implied in older medical textbooks. Growth stops because the cartilage cells of the growth-plate gradually respond less and less to growth hormone or somatomedin. Why they do this is not clear. It may be that the quantity or affinity of the receptors simply decreases (*see* Chapter 7) or there may be a more active process of repression of the receptors by some other substance.

Most limb bones have epiphyses at both ends. However, the femur grows mainly through the activity of the epiphysis at the knee, and little by activity at the hip. The humerus, in contrast, grows chiefly at the shoulder end. The forearm bones grow chiefly at the wrist, and the tibia and fibula about equally at knee and ankle. The vertebrae of the spine have growth-plates like those in limb bones, one on the top and one on the bottom. The small round bones of the wrist (carpals) and of the foot have growth-plates too, but they are concentric and surround the bone; thus these bones grow continuously from outside in towards their centre.

The growth in width of limb bones takes place in a different way, without the intervention of cartilage. The bone is covered by a sheet of fibrous material called the periosteum, which consists of two layers. The outer layer is truly fibrous, but the inner, known as the cambium by analogy with tree growth (where it is the layer between bark and wood) consists simply of cells which multiply to lay down new bone; in this respect they are like the cartilage cells in the proliferative layer of the growth-plate. Thus the bone grows in width by new layers of bone being deposited on the outside of the existing bone. Growth of the bones of the skull and of most of the bones of the face also takes place in this way.

A growing bone has to be continuously remodelled; the intricate shape of an adult femur cannot be obtained merely by adding length and breadth to the infant femur. Large areas of bone have to be continuously broken down, reabsorbed into the body and lost. The main features of a bone's shape are controlled by genetic instruction, for a tiny cartilage precursor of a limb bone can be grown in an artificial medium or under the skin of the back (in experiments using the mouse) and the bone will grow and ossify into a respectable imitation of its appearance when in its proper place. The final modelling, however, seems to depend on muscle pulls and stresses from the remainder of the skeleton. Certain bones are more plastic than others in development, and respond to stresses and strains of near-by muscles with considerable changes. This applies in particular to the alveolar margins of the upper and lower jaws, i.e. the bones directly under the gums, which bear the teeth. It is absolutely essential for the animal to have teeth which fit accurately together and jaws which are adapted one to the other. In the wild, failure in this respect means starvation. It seems that such accurate fitting was better achieved by plasticity and moulding than by rigorous genetic control of both sides of the bite. Thus nature plays into the hands of the orthodontists who can move teeth about and even alter somewhat the upper and lower jaw bones by applying pressures, in a way mostly denied to the childhood orthopaedic surgeon faced with the task of unbending deformed limbs.

CHAPTER 3

Growth Before Birth

The period of prenatal growth is vitally important to the child's future well-being; yet it is the period in which, inevitably, examination is hardest. Thanks to the development of ultrasound we now have a few longitudinal studies on prenatal growth, but during the first and second thirds of pregnancy we still have to rely for the most part on cross-sectional studies. Furthermore, in the second trimester we have to rely almost entirely on foetuses expelled from the uterus because one or the other was abnormal, whereas during the earliest weeks of pregnancy we have the mostly normal products of social abortions. For later foetal life we can study infants born prematurely, making the assumption that these children have grown before birth and will grow after it in exactly the same way as children who remain in the uterus the average length of time. Though this may seem a hazardous assumption, it is probably justified, if children with certain abnormalities are excluded.

Our ignorance begins at the beginning, for we do not know what forces are responsible for selecting out of millions the one sperm which fertilizes the ovum. Fertilization takes place in one of the tubes which lead from the ovaries to the uterus. The fertilized egg spends some six to seven days drifting down the tube and floating in the uterine cavity before it implants into the wall of the uterus. During this time the cells divide steadily, so that by the time of implantation the blastocyst, as it is called, consists of around 130 cells. Blastocysts can be washed out of the tube and implanted in foster-mothers, and in animals this technique has been used both for examining the effect of different kinds of uterine environment on development and for transporting by air the foetuses of large animals packed, temporarily, in the uteri of small ones.

After implantation the outer layer of the blastocyst undergoes a series of changes which culminate in the formation of the placenta. A small proportion of the inner layer develops into the embryo. The period of the embryo is considered to begin 2 weeks after fertilization and ends 8 weeks after fertilization, when the child, now recognizably human, with arms and legs, a heart that beats, and a nervous system that shows the beginnings of reflex responses to tactile stimuli, is called a foetus. At this stage it is about 3 cm long (*see* Figure 15).

This is a hazardous period, and many more ova are fertilized than come to fruition. Some 10% fail to implant and of those which implant and

Figure 15   Foetus at 13 postmentrual weeks
(From D. W. Smith, 1977, by kind permission)

become embryos about 50% are spontaneously aborted, usually without
the mother knowing anything has happened. Such abortion is due in most
cases to abnormalities of development, either of the embryo or of its
protective and nutritive surrounding structures. Abnormalities of the
chromosomes, for example (*see* p.52) are present in 5% to 10% of fertilized
ova but in only 0·5% of newborns. Thus 90% to 95% of all conceptuses
with these abnormalities are rejected, as early spontaneous abortions.
Lesser degrees of abnormality may result in a viable foetus and child, but
one which grows less than is normal. An example of this is the child with
the disorder called Down syndrome, caused by possession of an abnormal
number of chromosomes (*see* p.52).

The reckoning of age in the prenatal period presents a problem.
Traditionally, and because we have no better way, age tends to be counted
from the first day of the last menstrual period. This occurs on average 2
weeks before ovulation, but with a variation of plus or minus 6 days about
this. Thus the most frequent age at birth is 280 days or 40·0 weeks reck-
oned as 'postmenstrual age', but this represents only 38 wccks of true

foetal age. There are difficulties in individual cases, however; the interval from menstruation to fertilization varies from 8 to 20 days, and worse, menstrual bleeding or something like it, may continue in some women for 1 or even 2 months after fertilization. In studies of organ differentiation and development, postfertilization age is frequently used (see, for example, the next chapter).

Reliable growth curves for the length of the foetus are hard to come by, for the reasons given at the beginning of this chapter. There are good modern data covering 10 to 18 weeks postmenstrual age, and, for infants born prematurely, reasonable ones from about 20 weeks onwards. Figure 16 is necessarily somewhat diagrammatic; it shows the distance and velocity curves of body length, so far as they may be measured in prenatal life, and for the first year after birth. The position of peak velocity is uncertain and may be up to 6 weeks later than shown. The solid lines represent the actual length and length velocity; the interrupted lines the theoretical curve which might be supposed to occur if no uterine restriction took place in the last weeks of pregnancy, followed by a compensating catch-up after birth (see p.165).

In the embryonic period the velocity is not very great. During these first 2 months differentiation of the originally homogeneous whole into regions, such as head, arms and so forth occurs, as does histogenesis, the differentiation of cells into specialized tissues such as muscle and nerve. At the same time each region is moulded into a definite shape, by differential growth of cells or by cell migration. This process, known as morphogenesis, continues right up to adulthood and, indeed, in some parts of the body, into old age. But the major part of it is completed by the 8th postmenstrual week.

The high rate of growth of the foetus compared with that of the child is largely due to the fact that cells are still multiplying. The proportion of cells undergoing division in any tissue becomes progressively less as the foetus gets older, and it is generally thought that few, if any, new nerve cells (as distinct from the supporting neuroglia) and only a small proportion of new muscle cells appear after 30 postmenstrual weeks, by which time the velocity in linear dimensions is dropping sharply. The muscle and nerve cells of the foetus, however, differ considerably in appearance from those of the child or adult, for early in development there is little cytoplasm around the nuclei (see Figure 12, p.29). In foetal muscles there is a great deal of intercellular substance and a much higher proportion of water than in mature muscle. The later foetal and postnatal growth of muscle consists chiefly of building up the cytoplasm of the muscle cells; salts are incorp-

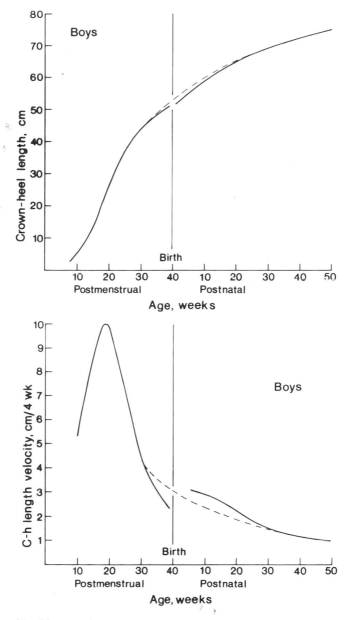

Figure 16 Distance (*above*) and velocity (*below*) curves for growth in body length in prenatal and early postnatal period. Diagrammatic, based on several sources of data. The solid lines represent actual length and length velocity; the interrupted line the theoretical curve if no uterine restriction took place
(Tanner, unpublished)

orated and the contractile proteins are formed. The cells become bigger, the intercellular substance largely disappears and the concentration of water decreases. This process continues quite actively up to about 3 years of age and slowly thereafter this; at adolescence it briefly speeds up again, particularly in boys, under the influence of androgenic (male-determining) hormones. In the foetal nerve cells cytoplasm is added, and the cell processes grow. Thus postnatal growth, for most tissues, is chiefly a period of development and enlargement of existing cells, whereas early foetal life is a period of cell division and the addition of new cells.

During the last 10 weeks in utero the foetus stores very considerable amounts of energy in the form of fat. Up till about 26 weeks postmenstrual age most of the increase in foetal weight is due to accumulation of protein as the main cells of the body are built up. During this period, however, the differentiation and laying down of the scaffolding of fat tissue begins. This occurs in a definite order. First the fat pad of the cheek can be identified, at about 14 weeks, then the tissue in neck and trunk (about 20 weeks) and then the lower limbs (about 22 weeks). By 25 weeks all the tissue seems to have been assembled, ready to receive the droplets that constitute the fat store (Poissonet et al., 1984).

From then on fat begins to accumulate, both deep in the body and subcutaneously. Drs D. A. T. Southgate and E. Hay (1976) have analysed 36 foetuses and found that from 30 to 40 postmenstrual weeks fat increases from some 30g to 430g. Since fat packages much more energy than protein or carbohydrate per unit volume, this represents a large reserve of energy available for the first, perhaps critical, period after birth (newborn monkeys do not store up fat in this way, hence look less cuddly to the human eye). Conversely, the creation of such a store represents a considerable drain on the energy resources of the mother in the last weeks of pregnancy.

The Effect of the Uterine Environment

Growth in weight of the foetus follows the same general pattern as growth in length, except that the peak velocity is reached later, at approximately 34 postmenstrual weeks. There is considerable evidence that, beginning at 34 to 36 weeks, the growth of the foetus slows down owing to the influence of the uterus, whose available space is by then becoming fully occupied. Twins slow down earlier, when their combined weight is approximately the 36-week weight of the singleton foetus. In Figure 17 is plotted a velocity curve constructed before birth from the differences

Figure 17   Velocity of growth in weight of singleton children. Prenatal curve drawn from data of Yudkin *et al.*, (1987) on birthweights of live-born male children delivered spontaneously before 40 weeks of gestation. Postnatal data from Ministry of Health (1959); mixed longitudinal data (their Table VII) (Tanner, unpublished)

between weight means of singleton male babies born spontaneously and alive at periods from 24 to 40 weeks, and after birth from differences in weight means in a mixed longitudinal study of a similar population in the UK. A curve comparable in shape with usual growth curves could be constructed by joining the top, 34-week prenatal point with the 12-week and subsequent postnatal points. Such a hypothetical curve would predicate a peak weight velocity reached at about 36 or 37 weeks. But between 34 and 40 weeks, growth is held up; and the increase in velocity in the first 12 weeks after birth represents a catch-up (*see* Chapter 10) on the part of those newborns who have been most delayed in the uterus. It is for this reason that there is significant negative correlation between length at birth and length gain in the first 6 months after birth, and also between weight at birth and weight gain in the first 6 months. The smaller the baby, on average the more it grows at this time.

This slowing-down mechanism enables a genetically large child developing in the uterus of a small mother to be successfully delivered. It operates in many species of animal; the most dramatic demonstration was made by crossing reciprocally a large Shire horse with a small Shetland pony. The pair in which the mother was the Shire had a large newborn foal, and the pair in which the mother was the Shetland had a small foal. But both foals were the same size after a few months, and both ended about half-way between their parents. The same has been shown to occur in cattle crosses.

The natural impulse to think of the prenatal cross-sectional curve of Figure 17 as representing the way individual foetuses grow should be resisted. There is the usual difficulty in arguing from cross-sectional to longitudinal data outlined above (p.10); but besides this, we would be assuming that the selection of a foetus for early spontaneous birth was independent of its size, and this might not be the case. If large foetuses were more often born early, then those appearing at 39 and 40 weeks would be predominantly small ones and this would cause a lowering of the mean-increment curve. If this were really the case however, one would not expect the small ones to catch up after birth, as they do, since they would not have deviated from their regular curve. However this may be, it is highly probable that the peak of weight velocity is reached at different times in different foetuses. There is some evidence, too, in the best data, that females reach their peak a week or so earlier than males, though this needs confirmation through the longitudinal data provided by ultrasound estimates of body weight.

A frequently used measure of the degree of association between two variables is called the *correlation coefficient*. It varies from 0·0, which means the two variables are quite independent, to 1·0, which means that one is totally predicted by the other. In man the correlation coefficient between length at birth and adult height is only about 0·3; but the coefficient rises sharply during the first year, and between length at age 2 and adult height is nearly 0·8. These figures also reflect maternal control of newborn size.

How this control is exercised is not clear. The placenta grows at first more rapidly than the foetus, but from about 30 weeks onwards its growth rate becomes less than that of the foetus and the placenta/foetus ratio falls. It may be that the placenta simply cannot increase its capacity to supply enough food to sustain the rapid 34-week foetal velocity. In mice and guinea pigs it seems likely that the limiting factor is blood flow, the size of the placenta depending on the pressure at which the maternal blood reaches it, and the size of the foetus depending, in turn, on the size of the placenta. Whether this is also important in man is not yet known though there is some evidence, collected by Dr A. Briend, that a fall-off in weight gain may be caused by the mother doing heavy work during the last few weeks of pregnancy, and particularly work which puts a strain on the haemodynamics of the uterus. Poor environmental circumstances, especially of nutrition, result in lowered birthweight in the human. This seems chiefly to be due to a reduced rate of growth in the last 2 to 4 weeks of foetal life, for mean weights of babies born at 36 weeks to 38 weeks in various parts of the world under various circumstances are rather similar. Mothers who have not achieved their full growth potential because of adverse circumstances in their own chidhood, may produce smaller foetuses than they would have done had they grown up under better conditions. Thus two generations or even more may be needed to undo the effect of poor environment on birthweight. In Guatemalan villages, for example, mothers of short stature had babies of lower birthweight than did mothers of medium stature. A food supplement given to the mothers during pregnancy had more effect on the birthweight of children born to short mothers than on those born to medium-sized mothers, but it did not wholly eliminate the difference. The present average weight at term of Swedish babies, who are the heaviest in the world, due to the excellence of social conditions and pre-natal care, is 3·65 kg for boys and 3·50 kg for girls.

So-called 'Premature' Babies

The average length of gestation, measured from the first day of the last menstrual period, is 280 days or 40 weeks. However, there is considerable individual variation around this figure (even when mistakes and inaccuracies in determining gestational age are set aside), and lengths of gestation from 259 days (37 completed weeks) to 293 days (42 completed weeks) are by international agreement regarded as normal. Babies born within these limits are called *term* babies. Babies born earlier are called *pre-term* babies and those born later *post-term* babies. Recent investigations lead to the conclusion that the limits of normal term should really have been set between 38 and 41 completed weeks instead of between 37 and 42 weeks.

Until a few years ago all babies who weighed less than 2,500 g (5·5 lb) at birth were designated 'premature' whatever their length of gestation or physiological state. This definition (promulaged by WHO in 1948) caused much confusion and has now been dropped; the word 'premature' has disappeared from scientific use. Babies less than 2,500 g at birth (WHO, 1961) are called 'low birthweight' babies; this low birthweight may be due either to their being born early, or to their being babies who are pathologically small for their length of gestation. The distinction is made by the use of standards such as the one shown in Figure 18, which gives the centiles for first-born girls only; children subsequent to the first are heavier (on average by 110 g) and boys are heavier than girls (on average by 150 g). The baby marked A weighed 2,500 g but was born a month early, at 36 weeks. She is at the 25th centile for first-born girls of this gestational age, thus perfectly normally grown. The baby B, also of 2,500 g, born at 40 weeks (full-term), is below the 5th centile and thus of very questionable normality. Such a baby is known as small-for-gestational age, or, colloquially, as 'small-for-dates'.

Since there is a tendency for some mothers always to have relatively small babies and other mothers relatively large ones, a more critical standard can be constructed in which the weight of the new baby is compared with the birthweight of his brothers and sisters (length of gestation being allowed for). To a large extent this family trait of size at

Figure 18   Standards for birthweight for gestational age, first-born girls: (*a*) 2,500 g, born at 36 weeks; (*b*) 2,500 g, born at 40 weeks
(From Tanner and Thomson, 1970)

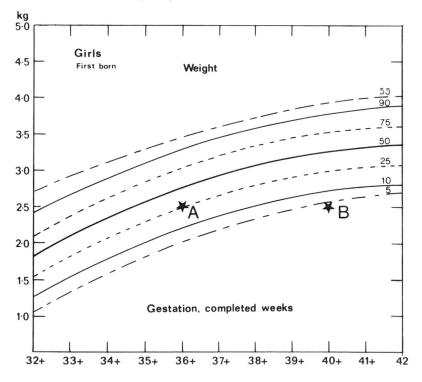

birth is inherited, probably through characteristics of the maternal uterus rather than of the foetus itself. This variation between families, however, remains strictly within the normal range of birthweights. When we talk about small-for-dates babies we mean those who are beyond these limits. A small overlap exists, for a very few normal babies are necessarily below the arbitrary limit set for the upper bound of small-for-dates infants. This occurs in the use of all statistical standards, as explained in the Introduction.

The distinction between pre-term babies of normal weight for length of gestation and babies who are light for their (often normal) length of gestation is an important one. Most of the purely pre-term infants (provided they are otherwise unimpaired) catch up perfectly well and seem little the worse for their earlier experience of the outside world. Even those born as early as 28 weeks weighing 1,000 g may nowadays be kept growing at the very rapid rate appropriate to their age and sent home 8 to 10 weeks later at the normal weight for a full-term infant. This follows the recent introduction of concentrated milk preparations together with intravenous nutritional supplements to provide the necessary amounts of calories and protein. Nilsen and his colleagues in Norway checked the heights at age 18 of 59 men who had birthweights below 2,500g. Those who, born early, had weights greater than the 10th centile for gestational age reached the same height as the national average, but those born later whose birthweight was below this centile averaged 6 cm less. (They were no lower in IQ, however.) These births took place around 1964, and there have been considerable improvements in treatment since then.

Babies who are light, that is, below the 3rd centile for length of gestation, do not on average catch up to normals, although they diminish the gap a little. The average small-for-dates child reaches about the 25th centile for height, which implies that a considerable proportion of such children remain below the 3rd centile limits of normal. A considerable proportion fail also to develop the same level of mental ability as normal children.

Children can of course be both pre-term and small-for-dates; about one-third of pre-term babies in the UK appear to be so. Such children fall clearly in the small-for-dates category. The deficits in later size and ability get worse as the birthweight decreases; children between 2,000 and 2,500g at (full-term) birth show little impairment of ability and only a slight size deficit. A considerable proportion of those under 2,000 g, however, have some mental or neurological defect. These figures concern, of course, survivors. The chance of perinatal death (i.e. death within 24 hours after birth) increases, the lower the weight for gestation. In a study of 44,000 consecutive births in the Royal Victoria Hospital in Montreal, Usher and McLean found that the perinatal mortality was 54 per 1,000 births in babies born at term (37 to 42 weeks) with weights under the 3rd centile for gestational age compared with 6 per 1,000 in babies with weights between the 30th and 70th centiles.

In babies of all lengths of gestation, perinatal mortality was 16 per 1,000 amongst babies whose weights were between the 3rd and 97th centiles for gestational age, but 189 per 1,000 for those whose weights were under the 3rd centile for gestational age.

Thus the prognosis for a small child born after a normal-length gestation is very different from the prognosis for an equally small child born after a shortened gestation. Leaving the uterus early is not in itself harmful, whereas growing less than normally during a full uterine stay implies pathology of foetus, placenta or mother.

Small-for-dates babies are a heterogeneous group, produced by several different causes. In countries where severe maternal malnutrition occurs a proportion of small-for-dates babies is due to this. But the malnutrition has to be fairly severe, for in this situation the foetus is protected at the expense of the mother. In well-nourished mothers disorders of the placenta may be responsible. Smoking in pregnancy causes, on average, a reduction of 180 g in full-term foetal weight and a 30% increase in perinatal mortality; the size reduction persists throughout childhood. The reduction is proportional to the number of cigarettes smoked per day, the value of 180g corresponding to 10/day. This reduction of weight brings some babies into the small-for-dates category. Smoking presumably works by affecting placental blood flow and foetal nutrition, though the less likely possibility of direct action on the foetal cells cannot be ruled out. Alcohol appears also to reduce foetal weight, and a large intake of alcohol may affect the foetus directly, causing a recognizable disorder known as the foetal alcohol syndrome. The face of the baby has a characteristic appearance, due to insufficient development of parts around the eyes, nose and upper lip. Some maternal diseases cause smallness-for-dates; in particular rubella (German measles), which also may cause specific deformities and deafness. In other instances it seems the foetus itself is disordered and lacks the capacity to grow properly. According to some studies, mothers of small-for-dates babies have a higher proportion of abnormal outcomes in other conceptions (miscarriages, etc.) than do mothers of normal-sized babies.

Since the worst period for the small-for-dates baby seems to be the last part of pregnancy, when he is, or should be, growing most, many doctors are now advocating removal of these foetuses at 36 weeks, or even at 34 weeks in severe cases. This is done by straightforward induction of labour and vaginal delivery. Much research is therefore in progress to devise means of recognising these foetuses in time for this action to be taken. To date, the most helpful guide to intra-uterine growth insufficiency is measurement of the foetus by ultrasonic means. Ultrasound may be used early in pregnancy to measure the length of the foetus's back from crown of head to rump, and from about 13 postmenstrual weeks to measure the width of the head and the circumference of the abdomen. It seems that screening all babies for crown-rump length between 6 and 12 weeks, for head width between 13 and 20 weeks, and for abdominal circumference at 32 weeks, would detect over 90% of all small-for-dates babies, with a negligible number of false positives (i.e. babies so diagnosed who were really normal). Such a screening programme is well within the capabilities of the medical services of industrialized countries. Biochemical tests for the integrity of placental function are being developed using the mother's blood or urine, but at the time of writing are not yet ready for general use.

Head Growth in Utero

Measurements of foetal weights have necessarily to be cross-sectional, but the advent of ultrasonic scanning has made possible fully longitudinal studies of the growth of the foetal head and abdomen. Particularly detailed studies have been made by Dr Russell Deter and his colleagues in Houston, who have fitted individual growth curves during fetal life in just the same way as we described in Chapter 1 for the postnatal period. Very-high-frequency sound waves (too high to be heard) are beamed at the foetus and echoed back from the junctions of differently conducting tissues. Thus the waves bounce back from the hard foetal head and the time taken for the echo to return indicates the distance from source to the edge of the skull. By scanning with a number of beams, and processing the results by computer, the foetal head may be outlined and accurate measurements of its maximal transverse diameter obtained without any harm being done. Foetal length from head to rump, and foetal abdominal circumference, can be estimated also.

Figure 19 shows two individual curves of head width from 20 to 40 weeks, from children of low birthweight, plotted against standards of head width for normal foetuses. Curve A shows a foetus with normal growth until 36 weeks; curve B a foetus whose head growth was abnormal from a very early age. Later follow-up of such children showed that those small-for-dates children whose head growth was normal, or whose velocity of head growth had only started to fall after 34 postmenstrual weeks (Figure

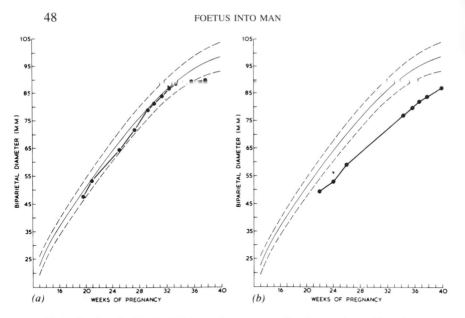

Figure 19 Growth of head width in two boys, measured by ultrasound scan. Chart shows mean and limits (3rd and 97th centiles) for each completed week of gestation: (*a*) normal growth till 34 weeks, then marked slowing; labour induced at 38 weeks, birthweight 2,450 g; (*b*) head already small at first measurement (22 weeks); slow growth began at 18 weeks or before. Birthweight, 1,890 g
(From Campbell (1976) and Fancourt *et al.*, (1976), with permission of authors and *British Medical Journal*)

19a), were normal in body size (or very close to it) and in intellectual development at follow-up. Those whose head-growth failure started before 34 weeks, and thereafter continued, were much smaller at follow-up, and had lower-than-normal intellectual attainment. Studies such as these are beginning to provide new insight into foetal growth and its pathology.

Birth as a Happening
For some physiological functions birth signifies upheaval and change, often associated with a particular vulnerability. But it is important to realize that for very many others birth is an incident without much significance in a steadily changing and maturing programme of events.

The respiratory and cardiovascular systems are the ones most altered by the fact of birth. Failure to establish satisfactory respiration during the crucial period just after birth is a common cause of neonatal death and of brain injury amongst survivors. The new-born infant, however, has a greater capacity to survive straightforward lack of oxygen without damage than have children or adults, and neonatologists are moving towards the view that in many cases it is pre-existing brain damage, of developmental or uterine origin, which is the reason for failure to start breathing. 'Almost all medical literature still equates failure to breathe at birth with "brain injury", placing the damage at the time of delivery. The evidence suggests that pre-existing brain damage interferes with the ability to adjust to extra-uterine life and to initiate respiration', wrote Drs H. Knobloch and B. Pasamanick (1962) in a review of mental subnormality. Dr C. M. Drillien (1964), whose study throughout childhood of babies in Scotland born small-for-dates is amongst the most thorough, remarks that the most severe defects originate usually at an early stage of foetal development rather than from damage to a potentially normal nervous system occurring in the last 3 months of pregnancy or during delivery. However, differences in later behaviour were found between children who had been subjected to certain obstetric hazards and those who had not, and these might have been wholly or partly due to minimal brain damage occurring at the time of delivery. The success or otherwise of the early delivery of small-for-dates babies will show which view is predominantly correct.

Some enzyme systems are sharply affected by the fact of birth. Thus in rabbits there is a sudden rise in glucose-6-phosphatase in the liver at birth and not before, whether birth is induced early, occurs normally, or is late. This is readily accounted for. The enzyme catalyses the breakdown of glycogen to glucose in the liver. Just after birth there is a great demand for glucose in the blood and tissues, pending the full establishment of lactation, and the enzymic events are adjusted in a way which satisfies this demand.

On the other hand, numerous enzymes exist with time courses of development which are independent of birth. A well-known example concerns the switch from production of foetal haemoglobin to adult haemoglobin. These are forms of haemoglobin with slightly different molecular structures. A gradual switch in production occurs, so that the percentage of foetal haemoglobin begins dropping from about the 36th postmenstrual week. The switch is not due to birth, however, because the percentage of foetal haemoglobin remains high in pre-term babies and is low in those who are born post-term. Thus in this, as in many other respects, the prematurely born has to a wait the striking of some previously regulated biological clock.

Most important, the maturation of the nervous system seems to be little affected by the fact of birth. The electroencephalogram (or 'brain wave' record) of the infant born at 28 weeks will 6 weeks later be much the same as that of an infant born at 34 weeks, provided adequate care is given. Dr C Dreyfuss-Brisac, the foremost expert in this field, wrote (1975): 'It is particularly important to point out that electroencephalographic maturation is closely and exclusively related to conceptional age . . . the duration of extra-uterine life is not an important factor in EEG maturation.' Birth also fails to influence the date of appearance of conditioned reflexes or the stages of motor behaviour elicited by detailed neurological examination. According to Dr St Anne Dargassies (1966) the development of motor behaviour occurs in a pre-term infant 'just as if birth before term did not perceptibly alter the course of neurological maturation. [The schedule] is adhered to just as closely in an incubator as in utero.' A strikingly simple example is given by the rates of conduction of nerve impulses along the peripheral nerves in the arm. The rate gets gradually faster up to about 2 years of age, when adult values are reached. The writer's colleagues Professor Dubowitz and Dr Moosa of the Institute of Child Health found that at 40 weeks after conception only very slight differences in conduction rate existed between normal-for-gestation pre-term babies born at 36 weeks and earlier, and normal babies born at 40 postmenstrual weeks. Five weeks later even this slight difference had disappeared.

More recently, Goeschen and colleagues in Berlin have shown that the conduction rate of the ulnar nerve motor fibres increases from 13·0 m/sec at 33 weeks postmenstrual to 30·4 m/sec at 44 weeks. The rate increased proportionately to age, irrespective of birth. Conduction rate is a good measure of biological maturity in the foetal period.

CHAPTER 4

Sex Differentiation up to Puberty

Chromosomal Sex

It is possible to take human white cells from the blood, or connective tissue cells from a tiny piece of skin, grow them outside the body and stop their activity at the moment in their division when the chromosomes are especially visible. The chromosomes, which are in the nucleus of the cell, are then stained and the cell is photographed. The chromosomes are the structures which carry the genetic blueprint, in DNA molecules (*see* p. 26) which specify whether the cell belongs to a man or a mouse, to a blue-eyed or brown-eyed man, a man who under favourable childhood circumstances will grow to be 6 ft tall or a man who under the same circumstances will grow to be 5 5 feet.

The chromosomes also specify whether the cell belongs to a male or a female. Chromosomes come in pairs; in a given cell of a child one chromosome of each pair is derived from the father and one from the mother. There are 23 pairs of chromosomes in man, and the members of each pair are distinguishable from members of all other pairs by their size, shape and staining patterns. When the chromosomes are stained in the cell nucleus they lie higgledy-piggledy and not paired up with their opposite numbers, so for clarity it is customary to take photographs, cut out each individual chromosome and then arrange them in pairs as shown in Figure 20. The chromosomal pattern thus displayed is called a *karyotype*. The members of each pair of chromosomes appear identical and indeed carry precisely corresponding pairs of genes, though in many of the pairs the two genes are in slightly different forms. There is one major exception, however; the members of one pair of chromosomes are grossly dissimilar. One, the X chromosome, is amongst the largest of the whole set, and the other, the Y chromosome, is amongst the smallest. These are the chromosomes that specify sex, for in all mammals creatures with two X chromsomes are females and those with one X and one Y are male.

This is where sex differences begin. Ova and sperm are special cells, and contain only one member of each chromosome pair. Thus when they join at fertilization the fertilized egg has the usual pairs of chromosomes, one member of each derived from each parent. Since the karyotype of the father is XY a sperm may carry either the X or the Y. The karyotype of the mother is XX so the ovum carries one or the other of these Xs. The sex of the child therefore depends on the sex of the sperm. When the fertilized egg divides, it passes a pair of chromosomes on to both daughter cells, which are thus XX from a female fertilized ovum or XY from a male fertilized ovum. The same occurs at all subsequent divisions (except those making new ova or sperm), so all the cells of the child have the karyotype XX if a girl, XY if a boy.

Sometimes this process goes wrong, usually because a pair of chromosomes fails to split cleanly into the singles needed for ovum and sperm. Mistakes may happen with any of the chromosomes but seem to be particularly common in the sex chromosomes. In as many as 1% to 2% of all conceptions a fertilized ovum is made with a single X chromosome only, without either a Y or a second X. The karyotype is written XO. The very great majority of such children, it is believed some 98%, die in utero, often so early in pregnancy that the mother does not even know she has aborted an embryo. This is as much or more due to failure of development of the placenta, whose cells of course suffer from the same chromosome anomaly, as the embryo itself. The few that survive (including mosaics, see below) constitute about one out of every 3,000 female babies. They are girls in external appearance, but their ovaries degenerate early in childhood so they cannot produce the female sex hormone at puberty which causes the breasts to grow. They are, of course, infertile. They are also strikingly short in stature, owing to a diminished response of the bones to the hormones controlling growth. The appearance of their face, neck, shoulders and chest is characteristic, though sometimes only distinguishable by the expert. The condition is known as gonadal dysgenesis or Turner syndrome. Sometimes the defect in the mechanism of chromosome duplication occurs at a later stage than initial fertilization. Then there may result a child with two lines of cells, some being XO descendants of the erroneous division, and some normal XX. Such children are called mosaics; in appearance and function they may be anywhere between the full Turner syndrome and normal girls. A much higher percentage of these mosaic individuals survives the foetal period and they constitute about a quarter of the surviving children with Turner syndrome.

Other sex chromosome errors also occur. Children with the karyotype XXY (Klinefelter syndrome) occur about once in every 1,000 male live births; they are boys with characteristically long arms and legs, small testes, reduced fertility and often a slight degree of mental handicap. Other chromosomal mistakes produce such karyotypes as XXX, XXXY and XYY. The last anomaly, which also has an incidence of about one per 1,000 male live births, produces particularly tall men. Boys with XXY and XYY karyotypes nearly all survive foetal life.

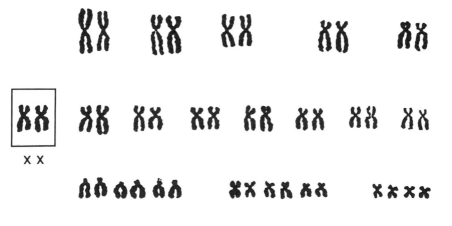

Figure 20 Karyotypes of normal male (*above*) and female (*below*)
(Reproduced by kind permission of Professor Paul Polani)

A similar mistake in chromosome No. 21 (*see* Figure 20), giving three instead of two 21s, is responsible for the well-known form of physical and mental handicap called Down syndrome. Chromosome 21 is small and presumably for this reason many triple-21 children survive; triplication of other chromosomes nearly always causes foetal death. It seems that these chromosomal errors occur more frequently in older mothers than in younger ones. The recent lowering of the average age of child-bearing in the UK has dramatically reduced the incidence of Down syndrome.

Prenatal Differentiation
During the earliest stages of development only examination of the chromosomes tells us whether the embryo is male or female. When the gonad (a term including both testes and ovaries) first appears, it looks the same in both sexes. Its rate of development, however, is greater in the male. This is due to a gene located on the short arm of the Y chromosome, which causes the secretion of a small protein called the testis determining factor. We do not yet know the intermediate steps by which this factor causes maleness. Between 7 and 8 weeks after fertilization the male gonad becomes recognizable as a testis; not until 2 weeks later does the female gonad undergo the changes which identify it as an ovary. About 9 weeks after fertilization, Leydig cells appear in the testes (*see* Chapter 7) and begin to secrete testosterone. The testosterone causes the previously undifferentiated external genitalia to form a penis and scrotum. This formation is complete by about 12 weeks after fertilization; thereafter the penis grows faster than the clitoris to produce the difference seen at birth. The stimulation of the Leydig cells during the period of formation of the genitalia is due to a hormone coming from the cells of the placenta (chorionic gonadotrophin). During the subsequent period of penis growth, testosterone levels are kept elevated by the same hormone plus perhaps some stimulation from the foetus's own pituitary gland. The testes also secrete a second hormone, from the Sertoli cells, which causes atrophy of the structures that in females develop into the Fallopian tubes, which carry eggs from the ovaries to the uterus.

In the absence of testosterone stimulation the external genitalia develop at around 12 weeks after fertilization into those of the female, apparently quite passively, without specific hormonal induction. It seems to be generally true throughout sex differentiation that the female is the 'basic' sex into which embryos develop if not stimulated to do otherwise. Even in the last stages of differentiation, at puberty, the male changes are more striking and extensive than those of the female, although in this instance girls as well as boys do have their own hormonal stimulus.

These changes lead to a qualitative difference – indeed to the only qualitative difference – between the sexes. Barring accidents of development the internal and external genitalia are always either male or female. Other sex differences – of body size, shape and tissue composition; of rate of maturing, strength, numbers of red blood cells and so on – are quantitative. Though a large difference in height, for example, exists between the average man and woman, there is overlap of the distributions, so that tall women are taller than short men.

The beginnings of many of these quantitative differences are traceable to the prenatal period. For example, the length of the forearm relative to the upper arm or to total body height is greater in the average male than in the average female. This sex difference begins in utero and continues very gradually to develop throughout the whole period of growth. The same is true of the intriguing and, from an evolutionary point of view, rather mysterious difference between the relative lengths of index and fourth fingers. The index finger is longer than the fourth more frequently in females than in males, and this pattern can already be seen at birth. (To make matters still odder, it is necessary to add that the right side of the body is on average more male in this matter, and indeed some others, than the left side; thus a man usually has a relatively longer fourth finger on the right hand and a relatively longer index on the left hand, a woman the reverse.)

Brain Differentiation

From an evolutionary point of view it is little use differentiating a peripheral organ such as the testis or ovary unless the messages causing them to function are differentiated also. The messages to the ovary cause it to operate cyclicly (hence the menstrual cycle); those to the testis cause it to operate non-cyclicly. These messages are hormonal and come from the pituitary gland (*see* p.87). But they are inititated and controlled by a portion of the brain called the hypothalamus which instructs the pituitary to send out its various hormones. Not only, then, has the endocrine system to be differentiated according to sex, but the hypothalamus also. It is known that in the rat, guinea pig, and many other species it is, as usual, the male brain which is changed, by testosterone secreted by the foetal testes. The change, involving cell structure, once made is irreversible. One can transplant a pituitary gland from an adult female rat into a male whose own gland has been removed and the transplant, when it starts to work, will obey the brain and function as male. In other words, endocrine maleness resides in the hypothalamus and not in the testis or the pituitary. Brain differentiation is a prime example of a sensitive period (*see* p. 172). In the rat, the testosterone must be secreted during the first 5 days after birth for the brain to differentiate. After this, it is too late; before this, it is too early and nothing happens. The rat is born relatively much earlier than the human, so this period corresponds to 12 to 14 postfertilization weeks (or 14 to 16 postmenstrual weeks) in man. It is known that blood testosterone level in the male foetus is high at this time but it is all the same uncertain whether differentiation exactly similar to that in the rat occurs in man. Probably it does, but some recent evidence in monkeys suggests that primates (monkeys, apes and man) may be somewhat different in the way they regulate their reproductive affairs, even at this level.

In the rat, there is evidence that a second, higher, brain centre, controlling types of mating behaviour, is also irreversibly differentiated. The control seems to be more quantitative than qualitative, which implies that only a partial reversal of sex role can be brought about by experimentally changing the animal's endocrine apparatus. It is at present unclear whether this second centre is present in man. If some such differentiation of the brain does occur, occasional errors in the process would certainly be expected. They would probably be quite rare, however, and could scarcely account for more than a small fraction of human sexual pathology.

Differentiation in Childhood
From birth till puberty sex differentiation continues, in the physical no less than in the psychological sense. The most important difference is in the rate of maturing.

Girls grow up faster than boys: that is, they reach 50% of their adult height at an earlier age (on average at 1·75 years compared with 2·0 years in boys), enter puberty earlier and cease earlier to grow. The difference in tempo of growth starts very early; half-way through the foetal period the skeleton is already some 3 weeks more advanced in girls than in boys. At birth the difference corresponds to 4 to 6 weeks of maturation and at the beginning of puberty to 2 years. Girls are physiologically more mature in some other organ systems too and it seems very likely that this is the reason why more girls than boys survive at birth, whatever the general level of perinatal mortality.

The actual steps in the maturing process, the milestones passed, are the same until puberty for girls and boys. They are also arranged in the same order of sequence along the road. But the intervals between some of them are different in the two sexes. The permanent teeth all erupt earlier in girls than in boys. However, some of them erupt much earlier, e.g. the canines with a difference of 11 months; others only slightly earlier, e.g. the first molars, with a difference of 2 months. Similarly, there are sex differences in the relative times of appearance of some centres of skeletal ossification, especially in the elbow region. Such differences must represent the small change of evolutionary advantage through some better co-operative arrangements for survival and for the rearing of young, but it is hard now to guess what they might have been. The earlier maturing of females is a characteristic shared by many mammals and nearly all primates so far examined. There are a few organs which constitute exceptions to the girls-earlier rule, most notably the milk teeth, which both sexes acquire at the same age.

Most of the sex differences in body shape and tissue composition are due to the events of puberty described in the next chapter. But the difference in overall body size is chiefly due to the boys' delay in growth. Though boys are slightly larger at birth than girls, the difference is small and remains so until the girls accelerate in their adolescent growth spurt. Girls then become temporarily taller. The boys, however, have meanwhile two years in which to continue their pre-adolescent growth before embarking on their own spurt. The actual amount of height gained in the spurt is, on average, greater in boys than in girls but only by some 3 to 5 cm. The height difference between men and women averages about 13 cm, of which 8 to 10 cm is due to those extra pre-pubertal years of growth in boys.

The relatively longer legs of men compared with women (relative, that is, to total height) are due to the same mechanism. During the immediately pre-pubertal years the legs are growing relatively faster than the trunk. Thus during the pre-adolescent years boys change proportions to become longer-legged, and though during the spurt itself growth in trunk length is slightly greater in boys, it is not sufficient to cancel the effect of the pre-pubertal growth.

CHAPTER 5

Puberty

Puberty is the time of the greatest sex differentiation since the early intra-uterine months. There are changes in the reproductive organs and the secondary sex characters, in body size and shape, in the relative proportions of muscle, fat and bone, and in a variety of physiological functions. In the older literature the word puberty was used to denote the appearance of pubic hair, but nowadays it refers to the period at which the testes, prostate gland and seminal vesicles, or the uterus and vagina, suddenly enlarge. In this way it is possible to refer to puberty in mammals in general; in the few who reproduce these changes at each breeding season, 'puberty' refers to the first of such occasions. (The word adolescence is increasingly used to refer exclusively to the psychological and behavioural changes occurring around this time. However, such phrases as 'adolescent growth spurt' are very much part of the literature of human growth, and 'puberty' and 'adolescence' will be used interchangeably in this chapter.)

Reproductive Organs
Males   Figure 21 shows the growth curve of the testes in healthy Swiss, Swedish and Dutch boys. The sizes of testes were determined by comparing them, by manual palpation, with standard models of increasing size. The models, for convenience assembled like a necklace, are known as the Prader orchidometer, after the originator, the professor of paediatrics in Zurich. (This is a regularly used clinical method, but different observers do at times give different ratings. The most accurate method is mapping by ultrasound.) Although the data are plotted against chronological age and thus subject to the difficulties described on page 11 above, they show the steep rise occurring at about 14 years and also the large differences between boys of the same age (as instanced by the gap between the 10th and 90th centile lines) that we shall discuss at length below.

Beginning testicular enlargement is usually the first sign of puberty in boys, accompanied by changes in the texture and colour of the skin of the scrotum. A little later the penis starts to enlarge and the pubic hair to appear. The sequence of events is shown in diagram form in the lower part of Figure 22. The solid areas marked *testis* and *penis* represent the periods of accelerated growth of these organs, and the horizontal lines and rating numbers marked *pubic hair* stand for its advent and development.

Testicular volume, ml (Orchidometer models)

Figure 21 Growth curve of testis size. Estimated 10th, 50th and 90th centiles, based on Dutch, Swiss and Swedish males (Data from van Wieringen *et al.*, 1971; Zachmann *et al.*, 1974; Taranger *et al.*, 1976)

The sequence and timings represent average values for European boys; North Americans average some four months earlier. To give an idea of the individual departures from average, the range of ages at which spurts for height, testis growth and penis growth begin and end are inserted underneath the first and last points of the curves or bars. The acceleration of penis growth, for example, begins on average at about age 12·5, but sometimes as early as 10·5 and sometimes as late as 14·5. The completion of penis development usually occurs at about age 14·5 but in some boys it is as early as 12·5 and in others as late as 16·5. There are a few boys therefore who do not begin this phase of pubertal development until the earliest maturers have completed it. Figure 23 (upper panel) shows three boys, all exactly the same age. At ages 13, 14 and 15 there is an enormous variability amongst any group of boys; this ineluctable fact of

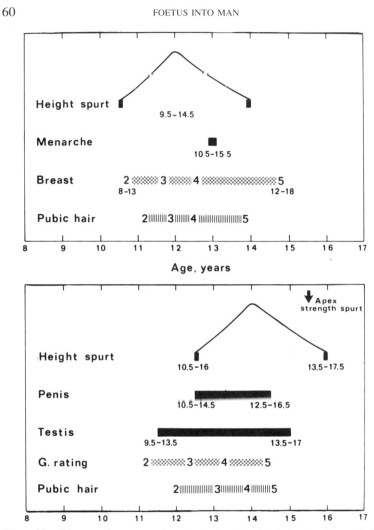

Figure 22   Diagram of sequence of events at puberty in girls (*above*) and boys (*below*)
(From Marshall and Tanner, 1970)

biology raises difficult social and educational problems, and contributes
to the psychological maladjustments often seen in adolescents.

The *sequence* of events, though not exactly the same for each boy, is
much less variable than the age at which the events occur. The spurt in
height and other body dimensions begins on average about a year after
the first testicular enlargement and reaches its maximum (peak height

Figure 23   Three boys, aged 14·0, at different stages of puberty; three girls, aged 12·0, at different stages of puberty
(From Tanner, 1975)

velocity) after about a further year, when the penis is also growing maximally and pubic hair is in stage 3 or 4. Axillary hair appears usually only when pubic hair has reached stage 4. Facial hair appears somewhat later, and usually in a definite order; first at the corners of the upper lip, then over the whole upper lip, then on the upper cheeks and the mid-line below the lower lip, and finally along the sides and lower border of the chin. The remainder of the body hair develops gradually from about the time of the first appearance of the moustache until a considerable time after puberty. The ultimate amount of general body hair that the individual develops seems largely to depend on heredity and is unrelated to the development of the specifically sexual hair on pubes and axilla.

Breaking of the voice happens relatively late in puberty and is due to the increased length of the vocal cords which follows the growth spurt of the larynx. However, since higher notes may still be made by correct stopping of the cords (as witness male counter-tenors), 'voice-breaking', which simply means the advent of lower tones due to the lengthening of the vocal cords, is often a gradual process and is not reliable as a criterion of puberty (though it was often used as such at periods when examination of genitalia and pubic hair was regarded as improper).

During adolescence the breast in boys undergoes changes, some temporary and some permanent. The diameter of the areola, which is equal in both sexes before puberty, increases considerably, though not as much as in girls. The average diameter before puberty is about 12·5 mm (0·5 in) and this approximately doubles in boys and triples in girls. In some boys (between one-fifth and one third of most groups studied) there is a distinct enlargement of the breast about midway through adolescence. When marked, this is known as gynaecomastia. The enlargement usually disappears spontaneously after about a year (if it does not, the tissue can be removed surgically).

Concurrently with growth of the penis, and under the same influence, the prostate gland and seminal vesicles enlarge. The secretion of the prostate makes up most of the seminal fluid. Thus no true ejaculation occurs before puberty, though the neural paths underlying orgasm are present. The first spontaneous ejaculation of seminal fluid usually occurs about a year after the beginning of penis growth, often during sleep and accompanied by appropriate dreaming. At first the seminal fluid seems to contain fewer and less viable sperm than in early adulthood; thus in males there may be an initial period of reduced fertility similar to that often occurring in girls (*see* below). In neither sex, however, can such relative infertility be relied on.

In clinical work some designation of how far a child has progressed through puberty is often required, and rating-scales have been developed for successive stages of growth of the penis and scrotum in the male, breasts in the female, and pubic hair in both sexes. The scales in each instance range from 1 (pre-pubescent) to 5 (adult). Details will be found in Tanner (1962), and illustrations in Chapter 11 (pp. 213 ff.). The size of the testes is assessed with the orchidometer. Genitalia and pubic hair do not necessarily develop in parallel.

There are large individual differences in the rapidity with which a given sequence, once begun, progresses to maturity. For example, on average boys take about a year to go from the beginning of stage G2 to the beginning of stage G3 and about 3 years to go from G2 to G5. But some boys take only two years to go from G2 all the way to G5. Thus out of a collection of boys all starting equal at the beginning of G2 a few reach G5 before others even reach G3. This variation in speed of development is additional to the variation in age of puberty as a whole, and for the most part independent of it.

Females   The same great variability is seen in the development of girls at puberty. The upper portion of Figure 22 shows the average age at which each event takes place, and the range of variation in healthy girls. Usually the first event to be noticed is the advent of breast-stage 2, called the breast bud. This consists of an elevation of breast and papilla as a small mound, with slight enlargement of the areolar area. The average age at which this occurs in European girls is about 10·8 years but the range extends from 8·8 to 12·8 (all such ranges given here refer to the mean plus and minus 2 standard deviations, which includes 95% of the normal population. Five per cent lie even beyond these limits). Usually pubic hair begins to appear some six months later; however, in about one-third of all girls pubic hair appears before the breast bud. The two series of events show considerable independence; thus among girls in breast-stage 3 all stages of pubic hair development are shown, 25% of girls having none at all, and 10% having reached full adult status. The size and structure of the uterus and ovaries can nowadays be ascertained by means of ultrasound. The uterus grows little before about age 9, but early in puberty it begins to enlarge, and by breast stage 3 it has doubled its childhood size. By stage 4 it is about four times the prepubertal volume and by stage 5, five times. The ovary, on the other hand, enlarges considerably during the prepubertal years, and at a fairly constant rate. Small ovarian follicles are present even in infancy and as growth proceeds larger and larger ones appear until by age 9 or so they look like cysts and begin to resemble the postpubertal follicle.

Menarche, the first menstrual period (pronouned menarkee, from the Greek word arçhē meaning beginning) occurs relatively late in puberty. The average age of its occurrence varies from 12·8 to 13·2 years in various present-day populations of northern and central Europe and in girls of European descent in North America. There are extensive data on age at menarche from all over the world, shown in Table 3 (p.146). The standard deviation of age at menarche is about 1 year in most populations so the 95% range is from about 11·0 years to about 15·0 years in West European girls. Again there is considerable independence from other pubertal characteristics. Most girls at menarche are in breast-stage 4 (projection of the nipple and areola above the level of the rest of the breast, in contrast to stage 5 in which only the nipple projects, the areola receding to the general breast level). However, some 25% are in breast-stage 3, and a small percentage in the earlier, breast-bud stage. Similarly most girls at menarche are in pubic-hair stages 3 or 4, but some are in stage 5 and a very few still in stage 1. The relation of menarche to the height spurt, however, is close. All girls start to menstruate when the height velocity is falling, i.e. during the downward part of the height velocity curve. Indeed, on average menarche occurs at the time of maximum decleration of height growth, the moment when the velocity is dropping fastest. The hormonal significance of this is unknown.

Though menarche marks a definitive and probably mature stage of uterine development, it does not signify the attainment of full reproductive function. The early menstrual cycles, which in some girls are more irregular than later ones, often take place without an egg being shed from the ovary. Thus there is frequently a period of adolescent sterility or partial sterility, lasting for 12 to 18 months after menarche. However, this does not occur in all girls.

There is an important difference between girls and boys in the relative positions of the height spurt in the whole sequence of development at puberty. It has only recently been realized, as a result of detailed longitudinal studies on relatively large numbers of children, that girls have their height spurt considerably earlier than boys. The difference between the ages of peak height velocity in girls and boys is just 2 years, but the difference in the first appearance of pubic hair is about 9 months. The first appearance of breasts precedes the first change in testis-size by even less. The adolescent growth spurt in girls is placed earlier in the sequence than in boys. Indeed in some girls peak height velocity is reached soon after breast-stage 2 and occasionally even before it. Thus for girls the first event of puberty is often an increase of growth velocity, seldom noticed. In boys, on the other hand, the peak height velocity very rarely occurs before stage 4 in genitalia growth is reached, and it is never the first sign of puberty. This has practical importance, for boys who are late maturers can be reassured that their height spurt is yet to come if genital development is not far advanced; and girls who worry about being too tall can be reassured their height spurt is nearly over if menarche has occurred. On average girls grow about 6 cm (roughly 2 in) after menarche, although gains of up to twice this amount may occur. The postmenarcheal gain is practically independent of whether menarche itself occurs early or late.

As in boys, some girls pass rapidly through all stages of breast or pubic hair development while others are slow. On average the time taken from first appearance of breast bud to reach breast-stage 3 is 1 year and to reach the adult stage is 4 years. Girls with rapid transit, however, take only 1·5 years to pass through all the stages, while slower developers take as much as 5 years or even more.

The Adolescent Growth Spurt: changes in body size and shape

At puberty there is a swift increase in body size, and a change in shape and in body composition. Though the changes are qualitatively similar for both boys and girls, they are quantitatively very different and their distinctive balance brings about much of the sexual dimorphism characteristic of adults. Thus before puberty males are only slightly taller than females whereas after it the difference averages 13 cm (5 in). At puberty the shoulders and muscles grow more in boys whereas the hips grow more in girls. Large differences in strength develop, together with other physiological distinctions. Some such sexual dimorphism occurs in all primates and indeed in most mammals, and relates to the behavioural as well as to the physical aspects of reproduction, and to the manner in which the species' social life is organized. The structural and physiological changes in boys make them more capable than girls of doing heavy physical work and of running faster and farther. Such sexual specializations are more notable in some species of primates than in others. Gibbons, for example, show little sex dimorphism; gorillas and rhesus monkeys much. Man lies about the middle of the primate range. Here again we have an ineluctable fact of biological history. The dimorphism may be no longer very relevant to many tasks in our push-button age but it is there nevertheless, with implications that cannot sensibly be denied or ignored.

The adolescent growth spurt in height is illustrated in the growth curves of the typical boy and girl given in Figure 5 (p.14). The rate of growth falls nearly continuously from birth onwards, and just before puberty it reaches its lowest point In the year just preceding his spurt the average boy grows about 5 0 cm and a gain of 3 5 cm is within the limits of normal. Thus boys with a delay of puberty who see others shooting up around them may feel they are not growing at all, so great is the contrast. Girls, with their earlier adolescent spurt, do not fall to quite so low a velocity before it. When the spurt starts, however, growth becomes very rapid. Typically a boy puts on about 7 cm in the first year of the spurt, 9 cm in the second and 7 cm in the third. After this, growth rapidly diminishes, with 3 cm in the next year and about 2 cm thereafter. In girls the velocity is lower, at about 6, 8 and 6 cm a year during the three years of the spurt. Naturally individuals differ much in the magnitude of their spurts; in Chapter 11 standard charts are given outlining the limits of normal. During the year which includes their peak velocities, however, most boys grow between 7 and 12 cm, and most girls between 6 and 11 cm. Children who have their peak early reach a higher peak than those who have it late; the early maturers are taking off from a higher level of velocity, though this may not be the only factor involved.

The average age at which the peak is reached depends on the nature and circumstances of the population, but in European and North American children at present it is about 12 0 years for girls and 14 0 years for boys. The variation around this is about two years either way; thus very early maturing boys reach their peak at 12 0 years, having started their spurt at about 10 5, while very late maturers begin the spurt only at 14 5 years, and reach their peak about 16 0. Corresponding variation occurs in girls.

The adolescent spurt is under somewhat different hormonal control from growth in the preceding period (*see* Chapter 7) and probably as a consequence the amount of height added during the spurt is to a considerable degree independent of the amount attained before. Most children who have grown steadily up the average curve have an average adolescent spurt, but some end up considerably above average as adults, and others considerably below. In statistical terms, the correlation coefficient between adult height and height immediately before the spurt starts is 0 8. This means that about 30% of the variability among men in adult height is caused by differences in the magnitude of their adolescent growth spurts. The same is true of women.

More of the spurt in height comes from acceleration in trunk length than from acceleration in growth of the legs (in practice, trunk length is measured by sitting height, which thus includes head height and pelvis height; and leg length is stature less sitting height). There is a fairly regular order in which the dimensions accelerate; leg length as a rule reaches its peak first, some 6 to 9 months ahead of trunk length. Shoulder and chest breadths are the last to reach their peaks. Thus a boy stops growing out of his trousers (at least in length) a year before he stops growing out of his jackets. At adolescence some children, particularly girls, complain of having large hands and feet; at least they can be reassured that by the time the spurt has ended their hands and feet will be a little smaller in proportion to arms, legs and height.

Nearly all skeletal and muscular dimensions take part in the spurt, though not to an equal degree. Even the head diameters, practically dormant since a few years after birth, show a small acceleration in most persons. The skull bones increase some 15% in thickness and there is growth in the tissues of the scalp. It is doubtful whether the brain itself has an adolescent growth spurt; if its does, the extent of it is small. The face undergoes considerable changes, illustrated in Figure 24. There is a clear adolescent spurt, coincident with the height spurt, in most facial dimensions, though not in the width of the eyes or the distance between them. The forehead becomes more prominent owing to growth of the brow ridges and the air sinuses behind them, and both jaws grow forward, the

Figure 24 Face changes in two boys at puberty
(From Tanner, 1962)

lower more than the upper. This change makes the profile straighter and the chin more pointed. At the same time the facial muscles develop. All these changes are considerably more marked in boys than in girls. Indeed in some girls scarcely any detectable spurt in face dimensions occurs at all. The greatest sexual dimorphism of the skeleton occurring at puberty, however, concerns the shoulders and hips. The curves are illustrated in Figure 25, aligned so that each individual's age is measured in years before and after his own peak height velocity. This minimizes the confusion caused by some individuals' having their spurt early and some late, and also places both sexes on the same time base. Girls have a particularly large adolescent spurt in hip width, a spurt which is quantitatively as great as that of the boys, despite the girls' spurt being a good deal less in nearly all other dimensions. The shoulder width spurt, on the other hand, is particularly marked in boys. These differences occur because cartilage cells in the hip joint are specialized to respond to female sex hormone (oestrogen) and cartilage cells in the shoulder region specialized to respond to male sex hormone (androgens, primarily testosterone). Such specializations can be lost as well as gained during the course of evolution. In most apes and monkeys the male has a large canine tooth, suitable for fighting. This normally develops at puberty, but it can be induced to grow before then by giving the animal testosterone. In man, the canine, though still slightly larger in the male, has apparently lost most of its ability to respond to androgens.

The shoulder–hip-width dimorphism has long been used as a measure of bodily androgyny, i.e. the degree to which the male resembles a female, or vice versa. In young adults an equation using just two measurements (known technically as a discriminant function) classifies correctly 90% of persons into male and female. The score which does this is *3 Biacromial diameter - 1 Bi-iliac diameter* and it may be used as a measure of androgyny which is less biased by sheer body size than is the simple ratio of shoulder to hip width.

Many sex dimorphisms antedate puberty, as pointed out in the preceding chapter. One of the most striking, the greater relative length of the male leg, comes about through males having a delayed adolescent spurt. Part of the sex difference in shape of the pelvis also antedates puberty. Girls at birth already have a wider pelvic outlet, i.e. a wider opening at the bottom of the bony pelvis, which constitutes the narrowest part of the passage through which the baby has to pass when it is born. Thus the adaptation for child-bearing is present from a very early age. The changes at puberty are concerned more with widening the pelvic inlet and broadening the

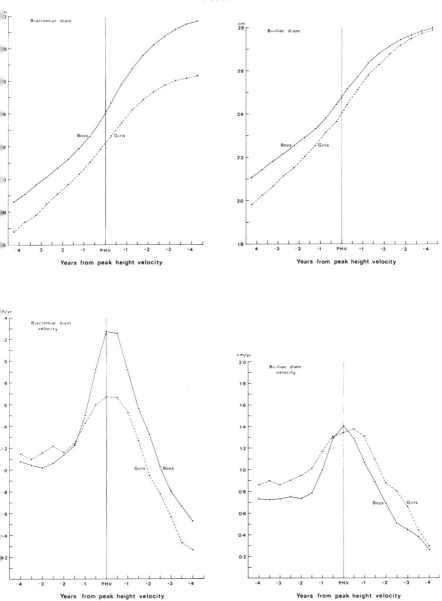

Figure 25 Biacromial and bi-iliac diameters at puberty: distance (*above*) and velocity (*below*) curves. Longitudinal data of Harpenden Growth Study plotted in years before and after peak velocity for each individual (From Tanner, 1975)

hips. This is important in providing room for the uterus and baby to grow
in.

Perhaps the hip widening may also have the subsidiary function of
attracting the male's attention: for these sex-differentiated morphological
characters arising at puberty – to which we may add the corresponding
physiological and possibly psychological ones as well – are developed not
only for immediate use in copulation, but for all the range of behaviour that
accompanies sexual reproduction. The wide shoulders and muscular
power of the male may have been developed partly for use in mating, but
more probably, with the canine teeth and prominent brow ridges, for
driving away other males as well as for hunting. The dimorphism is at first
maintained by sexual selection, since fighting involves males of the same
species. When this type of sexual selection becomes less important, as in
man, genetical theory leads us to suppose that these traits are gradually
eliminated.

But a number persist, through a mechanism known as ritualization. In
the course of evolution a morphological character or a piece of behaviour
may lose its original function and, becoming further elaborated, complicated or simplified, may serve as a sign-stimulus to other members of
the same group, releasing behaviour that is in some way advantageous to
the spread or survival of the species. It requires little insight into human
erotics to suppose that the shoulders, the hips and buttocks, and the breasts
(at least in a number of widespread cultures) serve as releasers of mating
behaviour. The pubic hair (about whose function the medical textbooks
have always preserved a cautious silence) probably survives as a ritualized
stimulus for sexual activity, developed by simplification from the hair
remaining in the inguinal and axillary regions for the infant to cling to
when still transported, as in many apes and monkeys, under the mother's
body. Similar considerations may apply to the axillary hair, which is
associated with special glands which themselves only develop at puberty
and are related in structure to scent glands in other mammals. The beard,
on the other hand, may still be more frightening to other males than
enticing to females. But perhaps there are two sorts of beard.

The Adolescent Growth Spurt: changes in proportions of muscle, fat and bone

There are marked changes in body composition accompanying the
adolescent spurt, especially in boys. Figure 26 shows the increase of limb
muscle, estimated from radiographs of arm and calf taken so that bone,

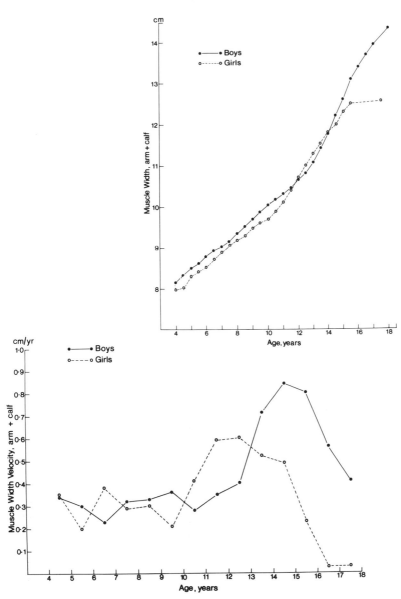

Figure 26   Muscle growth at adolescence: sum of widths of upper arm and calf muscles as visualized in X-ray pictures: (*above*) muscle width attained; (*below*) velocity of muscle width
(From Tanner, Hughes and Whitehouse, unpublished)

muscle and fat can be distinguished. The peak of the muscle growth velocity is somewhat later than the peak height velocity and thus occurs coincidentally with the peak of sitting height and shoulder width. It is much higher in boys than in girls. However, girls do have an adolescent spurt in muscle, and since it occurs earlier than in boys, girls on average actually have more muscle than boys for a short period, as Figure 26 shows. The increase in size of muscles reflects increased amounts of contractile protein and of nuclei, and underlies the great increase of strength occurring simultaneously (*see below*). Other muscles, including the heart, show similar growth curves.

Subcutaneous fat, or at least subcutaneous fat on the limbs, has quite the opposite curve. Figure 27 shows the velocity curve of limb fat in a group of children followed longitudinally, whose curves have been aligned as before, so that age is measured in years before and after their own peak height velocity. In boys the area of fat seen in the X-rays

Figure 27   Subcutaneous fat growth in upper arm and calf at adolescence, as visualized in X-ray pictures
(From Tanner, 1975)

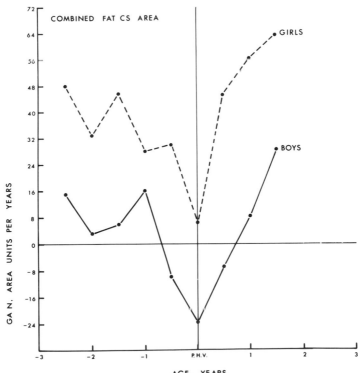

diminishes markedly, and the velocity actually becomes negative, indicating absolute loss. The lowest point is coincident with the peak of height velocity. In girls a similar thing occurs but is less marked, the decrease not being sufficient to carry the mean below zero. Thus no actual loss occurs in the average girl, who has to content herself with a temporary slowing down of fat accumulation. Fat on the chest and abdomen shows a much smaller decrease of velocity than fat on the limbs.

Strength, Exercise Tolerance and Other Physiological Functions
The increase in muscle size is naturally accompanied by an increase in strength, illustrated in Figure 28. This is much greater in boys than in girls, and it seems probable that the strength per gramme of muscle becomes greater in boys at this time, owing to changes in the structural and biochemical nature of the muscle cells induced by male sex hormone. The curves in Figure 28 are from a longitudinal study. Each individual test represents the best of three trials, made with each child competing against

Figure 28 Strength of arm pull and arm thrust from ages 11 to 17. Mixed longitudinal data, 65–93 boys and 66–93 girls in each group
(From Tanner, 1962, based on data of Jones)

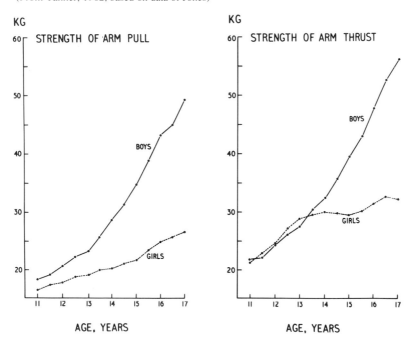

his own previous reading, and also against a classmate of similar ability. Only with such precautions can maximal values be approached. Other data indicate that except in hands and forearms, girls and boys are similar in strength for given body size until puberty begins, when a wide gap opens even between persons of the same overall size.

Boys develop larger hearts as well as larger skeletal muscles, larger lungs, higher systolic blood pressure, lower resting heart-rate, a greater capacity for carrying oxygen in the blood, and a greater power of neutralizing the chemical products of muscular exercise. The number of red blood cells and the amount of haemoglobin, the pigment which carries oxygen from lungs to muscles, both increase at adolescence in boys, but remain unchanged in girls. In short, the male becomes more adapted at puberty for the tasks of hunting, fighting and manipulating all sorts of heavy objects.

It is as a direct result of these changes that athletic ability increases so much in boys at adolescence. In a study by Dr H. C. G. Kemper in Amsterdam ordinary teenagers were induced to exercise on a treadmill to their very greatest capacity, when they were using oxygen at their greatest possible rate.During puberty there was an increase in this rate in boys from 2·2 litres/minute 2 years before peak height velocity to 3·7 litres/minute 2 years after peak height velocity. The girls' rise was smaller, from 1·9 to 2·6 litres/minute. The popular notion of a boy 'outgrowing his strength' at puberty has little scientific support. On the contrary, power, athletic skill and physical endurance all increase progressively and rapidly throughout adolescence. If the adolescent becomes weak and easily exhausted, it is usually for psychological reasons, not physiological ones. However, there is a brief period during which trunk length has increased relatively to the legs, bringing new problems of balance, while the muscles have yet to reach their full size and strength. This period seldom lasts more than six months, but may bring temporary problems in such specialized groups as young male ballet dancers, weight-lifters and field-event athletes.

CHAPTER 6

Developmental Age, and the Problems of Early and Late Maturers

Puberty starts at very different ages in different individuals, even of the same sex. In the last chapter, in Figure 23, p.61, we saw how perfectly normal 14 -year-old boys or 12-year-old girls could be pre-pubertal, mid-pubertal or post-pubertal. Manifestly it is ridiculous to consider the three boys illustrated as equally grown up physically, psychologically or – since much behaviour at this age is conditioned by physical status – in their social relations. One simply should not talk of '14-year-olds': the statement that a boy is 14 is hopelessly vague, for so much depends on whether he is an early or a late maturer.

These differences in rate of maturation do not suddenly arise at adolescence, although they become more obvious then. They are present at all ages, even at birth. Just as the average girl is developmentally ahead of the average boy from the middle of the foetal period onwards, so the early-maturing boy, on average, is always ahead of the late-maturing. This tendency for development to be rapid or slow was aptly named *tempo of growth* by Franz Boas, one of the great pioneers of human auxology. The analogy comes from classical music; some children play out their growth *andante*, others *allegro*, a few *lentissimo*. It seems that heredity plays a large part in setting the metronome, but we do not know the physiological mechanism.

Evidently we need a measure of developmental age, or physiological maturity, which represents more truthfully than chronological age how far a given individual has progressed along his or her road to full maturity. Height is not such a measure, because persons differ in mature height. Thus tallness in a child may signify either a rapid tempo of growth in a child going to be of average height when adult, or an average tempo of growth in a child going to be tall when adult. The percentage of the individual's own mature height reached at any age is a better measure, but available only retrospectively. Age at entry to the various stages of puberty is a valid measure, since all normal children pass through the same stages; but this measure also is available only late in growth. However, there are many possible measures of developmental age, ranging from the number of erupted teeth to the percentage of water in muscle cells. The various 'age' scales do not necessarily coincide and each

75

has its particular use. The measure most generally used derives from the successive stages of development of the skeleton as seen in radiographs. This measure is applicable throughout the whole period of growth and is called *skeletal maturity* or *bone age*.

Skeletal maturity is a measure of how far the bones of an area have progressed towards maturity, not in size, but in shape and in their relative positions one to another, as visualized in a radiograph. Each bone begins as a primary centre of ossification, passes through various stages of enlargement and shaping of the ossified area, acquires, in many cases, one or more epiphyses, i.e. other centres whose ossification begins independently of the main centre, and finally reaches adult form when these epiphyses fuse with the main body of the bone (*see* p.32). All these changes can be easily seen in radiographs, which distinguish the ossified area, whose calcium content renders it opaque to the X-rays. The sequence of changes of shape through which each bone passes is the same in all individuals, and all individuals reach the same final state.

In principle, any or all parts of the skeleton could be used, but in practice the hand and wrist forms the most convenient area. It is easily X-rayed without fear of any radiation being delivered to the reproductive organs, and it requires only a minute dose of X-rays, of the order of 4 millirems. There is no evidence that such a dose does any harm. Indeed it would be surprising if it did, since it represents the dose obtained unavoidably by every child who spends a week on holiday in the mountains (background radiation is 100 millirems per year at sea level, rising to 300 millirems at 2,000 metres altitude). The public is justifiably worried about the medical and genetic effects of atomic fallout and about the doses of radiation given in some forms of diagnostic radiology, but it is necessary to think quantitatively to keep matters in perspective.

Figure 29 shows two radiographs, both of boys aged 14·0 years. The one on the left has many epiphyses still unfused, and the wrist bones are not yet fully shaped. The hand on the right is approaching maturity. Skeletal maturity is assessed either by the 'atlas' bone age method of Greulich and Pyle or by the more detailed and perhaps more accurate bone-scoring method of Tanner and Whitehouse. In the former each bone is matched with a similar-appearing bone in a series of standard radiographs of increasing bone age. Thus each bone has a bone age assigned to it, and the modal (or most frequent) of these bone ages is taken as the bone age of the hand and wrist. Frequently the step of assigning bone ages to each separate bone is omitted; instead, the overall appearance of the given radiograph is simply compared with the appearance of the standard

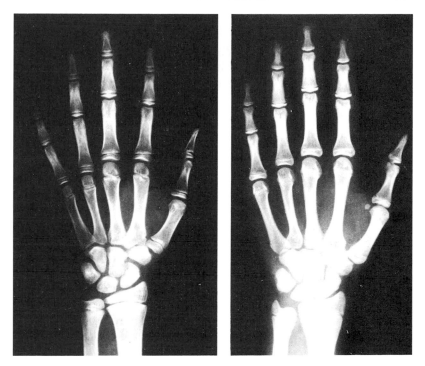

Figure 29 Radiographs of two boys both aged 14·0 years: (*left*) bone age 12·0 'years';
(*right*) bone age 16·0 'years'

radiograph, and the nearest overall match is taken. But such short-circuiting all too easily leads to inaccurate work.

In the Tanner-Whitehouse method each bone in the given radiograph is matched with a set of 8 standard stages and its stage is rated in just the same way as the puberty stages. Short-circuiting is impossible. Each stage of each bone has a score attached to it, derived mathematically so that the sum of the scores for all the bones represents the best overall estimate of skeletal maturity. In this way, a score for skeletal maturity is obtained which is comparable with a measure of height except that it runs from 1 to 1000. The maturity score may be turned into a bone age if desired, the bone age being simply the chronological age for which the given score is at the 50th centile. The Tanner-Whitehouse standards are based on a large-scale random sample of Scottish urban and rural children taken in the 1950s, and the statement that a certain boy has a bone age of 8·0 'years' means that he had the same degree of skeletal maturity as the average 8·0-

year-old Scottish boy of the area, social class and time. The Greulich-Pyle standards are based on a smaller number of very well-off North American children in the 1930s, who were some 6 to 9 months ahead of the Scottish children in maturity. Thus the two methods do not lead to the same result unless a correction is made. Workers in different countries or geographical areas have to establish local norms for skeletal maturity just as they do for height (see Chapter 9).

Boys and girls pass through identical stages in the Tanner-Whitehouse system and hence are rated in the same way. However, the scores attached to each of the stages are somewhat different, reflecting a sex difference in manner of maturation. The total maturity-score, obtained by summing the separate stage scores, is, of course, much higher in girls than in boys of the same chronological age. Girls are advanced compared with boys from before birth till maturity, the bone age of boys averaging about 80% that of girls.

The hand and wrist provides useful information only from about 18 months onwards, so other parts of the skeleton, chiefly the legs and feet, have to be used from birth to about 18 months. For some purposes not all the bones of the hand and wrist need to be scored; the forearm bones (radius and ulna) and the finger bones (metacarpals and phalanges) suffice. This is the case when bone age is used to predict the adult height that a child will reach. At a given chronological age an advanced child will be nearer adult height than a delayed child, and allowing for this makes prediction more accurate. Tables for prediction, with and without the use of bone age, are given in *Assessment of Skeletal Maturity and Prediction of Adult Height* (see Tanner *et al.*, 1983). Dr Alex Roche, at the Fels Research Institute, together with statistical collaborators, has provided a method for assessing the knee for skeletal maturity, and shown that hand and knee may not always give a closely similar result, even in healthy children.

Dental Maturity

The age of eruption of the teeth may also be taken as a measure of maturity. The deciduous dentition erupts from 6 months to 2 years of age and can be used during that period. The permanent dentition provides a measure from about 6 to 13 years. From ages 2 to 6 and from 13 onwards little information is obtainable from simple counting of the number of teeth erupted, but stages of calcification of teeth as seen in jaw radiographs can be used for dental maturity in exactly the same way as stages of the bones of the hand and wrist (*see* Demerjian *et al.*, 1973).

Relations Between Different Measures of Maturity

Skeletal maturity and dental maturity are not closely related. It is true that from 6 to 13 years children who are advanced skeletally have on average more erupted teeth than those who are skeletally delayed. Likewise children who enter puberty early erupt their teeth early on average, and girls have on average more erupted teeth than boys. Such average relationships, however, have little application to the individual. The correlation between bone age and dental age at a given chronological age is of the order of 0·4. This means that the two systems, though reflecting faintly the underlying presence of an overall general factor of maturity, measure to a great extent separate sorts of maturation. This is scarcely surprising, considering how very different are the timings of the growth curves of the head and its associated structures and the growth of the rest of the skeleton.

Skeletal maturity is more closey related to the age at which puberty occurs, though even this relationship is less close than was formerly thought, at least within the range of normal children. The milestones that are most closely linked are age at menarche and bone age. Figure 30 shows the frequency distribution of age of menarche in relation to chronological

Figure 30   Distribution of age of menarche in terms of chronological age and skeletal age. Girls of Harpenden Growth Study (From Marshall, 1974)

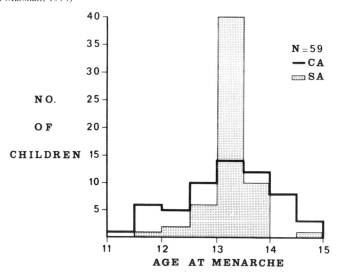

age and skeletal age in the same, normal, girls. Evidently the variation in relation to bone age is much less; indeed the 95% ranges are 11·0 to 15·0 in terms of chronological age but only 12·0 to 14·0 in terms of bone age. The link between skeletal maturity and the beginnings of puberty (i.e. the appearance of breasts or the enlargement of testes) is looser. However, at the extremes of advancement or delay in normal children, and even more in the abnormally advanced or delayed, skeletal maturity is a good guide to what is happening and what is likely to happen. If a boy is very much delayed in puberty then estimation of his bone age provides important information. If bone age is delayed then development is all of a piece and we may expect puberty to occur in the course of time; he simply has delay in growth. If the bone age is at an immediately pre-pubertal level of about 14·0 'years' and remains so over a period, then something is actually stopping puberty from occurring. The way in which the initiation of puberty is controlled is discussed in the next chapter.

Early and Late Maturing, Mental Ability and Emotional Development

Clearly the occurrence of differences of tempo in growth may have profound implications for educational theory and practice, especially if advancement in physical growth is linked to any significant degree with advancement in intellectual ability and emotional maturity.

There is good evidence that, in European and North American school systems, children who are physically advanced towards maturity score on average slightly higher in most tests of mental ability than children of the same age who are physically less mature. The difference is not great, but it is consistent, and it occurs at all ages that have been studied, going back as far as 6·5 years. Similarly, the intelligence test score of postmenarcheal girls is higher than the score of premenarcheal girls of the same age. Thus in age-linked examinations physically fast-maturing children have a significantly better chance than slow-maturing children.

It is also true that physically large children score higher than small ones, at all ages from 6 years on. In a random sample of all Scottish 11-year-old children, comprising 6,940 pupils, the correlation between height and score in the Moray House group test of ability was 0·25 ± 0·01, allowing for the effect of age differences within the range 11·0 to 11·9 years. An approximate conversion of these scores to Terman-Merrill IQ leads to an average increase of 0·67 points for each centimetre of stature, or roughly 1·5 points per inch. A similar correlation has been found in London children. The effects can be very significant for individual children. In 10-year-old girls there was a 9-point difference in IQ between those whose height was above the 75th centile and those whose height was below the 15th. This is two-thirds of the standard deviation of the test score. Children with many siblings (brothers and sisters) are shorter in stature and score less in intelligence tests than children with few sibs. About half of the correlation in the paragraph above is associated with differences in numbers of sibs; but about half remains when the number of sibs is allowed for.

It used to be thought that both the relationship between test score and height and between test score and early maturing would disappear in adulthood. If the correlations represented only the effects of co-advancement both of mental ability and physical growth, this would be expected to happen. There is indeed no difference in height between early- and late-maturing boys when both have finished growing. But it is now clear that, curiously, at least part of the height-IQ correlation persists in adults. It is not clear in what proportion genetical and environmental factors are responsible; differential social mobility (i.e. children's rising or falling in the social hierarchy relative to their parents) is probably the main factor involved (*see also* Chapter 9, p.153).

There is little doubt that being an early or a late maturer may have repercussions on behaviour and that in some children these repercussions may be considerable. There is little enough solid information on the relation between emotional and physiological development, but what there is supports the commonsense notion that emotional attitudes are clearly related to physiological events.

The world of the small boy is one where physical prowess brings prestige as well as success, where the body is very much an instrument of the person. Boys who are advanced in development, not only at puberty but before as well, are more likely than others to be leaders. Indeed this is reinforced by the fact that muscular, powerful boys on average mature earlier than others and have an early adolescent growth spurt. The athletically built boy not only tends to dominate his fellows before puberty, but by getting an early start is in a good position to continue that domination. The unathletic, lanky boy, unable, perhaps, to hold his own in the pre-adolescent rough and tumble, gets still further pushed to the wall at adolescence as he sees others shoot up while he remains nearly stationary in growth. Even boys several years younger now suddenly surpass him in size and athletic skill and perhaps too in social grace. Figure 31 shows the height curves of two boys, the first an early-maturing,

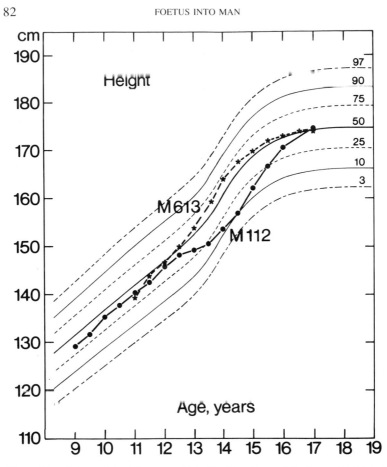

Figure 31   Height attained from ages 11 to 17 of two boys of the Harpenden Growth
Study. One (M 613) has an early, and the other (M 112) a later, adolescent spurt. The plots
are made against the Tanner-Whitehouse 1975 British Standards
(Redrawn from Tanner, 1977)

muscular boy, the other a late-maturing, lanky boy. Although both boys
are of average height at age 11, and then are together again at average
height at age 17, the early maturer is 12 cm taller during the peak of
adolescence.

The late developing girl, though not perhaps so plagued by athletic
problems, may have particular difficulty in her social relations because
her lack of breasts is plain for all to see. Indeed, late developers at
adolescence may sometimes have doubts about whether they will ever
develop their bodies properly and whether they will be as well endowed

sexually as others seen developing around them. At a deeper level the lack of development may act as a trigger to reverberate fears accumulated deeper in the mind during the early years of life.

It may seem as though early maturers have things all their own way. But early maturers have their difficulties also, particularly the girls in some societies. Though some glory in their new possessions, others are embarrassed by them. The early maturer has a longer period of frustration of sex drive, and of drive towards independence and the establishment of vocational orientation. Early-maturing girls are on average fatter than late-maturing, before, during and after puberty (just as early maturing boys are more muscular). Analysing the date of a very large collaborative study, Dr Stanley Garn, of the University of Michigan, showed that at age 30 women who had had menarche before age 11·0, had 30% more fat than women with menarche after age 14·0.

Studies by Dr Paul Mussen at the University of California have given evidence of the psychological consequences of early and late maturing. Late-maturing boys showed more attention getting behaviour and were rated by their age peers and trained observers as more restless, talkative and bossy. They were less popular and had lower social status. In contrast, the outstanding leaders came from the early-maturing group. Projective techniques revealed in the late maturers deeper feelings of inadequacy, greater anxiety and greater anticipation of rejection by the group. Follow-up studies in young adulthood showed that, at least in the society of the United States, the early-maturing boys became more sociable and less neurotic.

Little can be done to diminish the individual differences in children's tempo of growth, for they are biologically rooted and not significantly reducible by any social action. It therefore behoves all teachers, psychologists and doctors to be both fully aware of the facts and alert to the individual problems they raise.

CHAPTER 7

The Endocrinology of Growth

The endocrine glands are collections of cells which manufacture chemical agents and discharge them into the blood-stream. The agents are known as hormomes. Strictly speaking, a hormone is any substance secreted by one cell which passes to and has an action on another. Some hormones act on cells very close to their origin and do not enter the blood; some enter the blood but only in a special portion of the circulation designed to carry them a short distance to their target; and some enter the general blood-stream and are carried in it all over the body. Thus hormones constitute one of the two great systems of communication in the body (the nervous system being the other). This is as true for the carrying over of messages from one time to another as it is for the carrying of messages from place to place. The endocrine system is one of the chief agents for translating the instructions of the genes into the reality of the adult form, at the pace, and with the result, permitted by the available environment.

The hormones that are released into the general circulation pass by all manner of cells, but their targets are highly specific. This specificity is brought about by the existence of particular receptors in the target cells. The receptors are protein molecules; each abstracts its particular hormone from the blood, and links to it. Some receptor molecules are situated on the outer membrane of the cell, and others within the cell, in the cytoplasm. Those on the membranes react with hormones which are too large to be able to diffuse freely in and out of cells, as small molecules can. These hormones have to be grabbed from the blood (or more accurately from the tissue fluid, a filtrate of blood) as it flows by the cells. They are mostly peptides (strings of amino-acids linked together, in the same way as in proteins, which are essentially just very large peptides). The hormones from the pituitary gland (*see* p.87) are peptides of this sort. When the hormone and receptor link up on the membrane an intracellular messenger is released which speeds up a particular piece of cell machinery, in many cases by entering the cell nucleus and acting on the DNA of the chromosomes themselves. For many of these different hormone-receptor complexes the intracellular messenger is the same substance, called cyclic adenosinemonophosphate (cyclic AMP).

Smaller hormones pass back and forth through the cell membranes as a matter of course and are to be found in the cytoplasm of virtually all cells in the body, in concentrations reflecting their current levels in the blood.

Their presence has no effect except on the particular cells which contain receptors. Receptors for these hormones are placed in the cytoplasm, and the link-up occurs there. The linked complex then itself travels into the nucleus and turns on a particular part of the DNA mechanism for manufacturing protein. Even the fine details of this intricate process are now becoming clear (see O'Malley and Schrader, 1976). Steroid hormones, which include testosterone and oestradiol as well as cortisol (see below) act in this way. Thus in the muscle cells of the uterus there are many receptors for oestradiol, the female sex hormone. When blood oestradiol rises, the receptor picks up more molecules and the complex causes new muscle protein to be made: the uterus grows.

Thus receptors are as important in endocrinology as the hormones themselves. Indeed it seems that the evolution of this system of communication is more a story of receptor development than of changes in the forms of the hormones, though these have occurred, too, as species evolved (see, for example, growth hormone, p.89). A particular receptor may of course be present in all sorts of different tissues scattered all over the body; hence the ability of testosterone to act on cells in the penis, muscles, shoulder cartilage and brain. Different tissues contain different amounts. It has been calculated that 80% of cells in the uterus (at least of rats) contain receptors for oestradiol, at about 12,000 molecules per cell, while in the hypothalamus only half the cells have oestrogen receptors, and at the lower level of 4,000 molecules per cell. The hormone-receptor mechanism is extremely sensitive; O'Malley and Schrader use a striking simile: 'If the palate were as sensitive to flavours as the target cells are to hormone', they write, 'we would be able to detect a pinch of sugar dissolved in a swimming pool.' Receptors allow for a certain complication of action also. Hormone A may act by altering the sensitivity of receptor B[1] to hormone B. If it does so it will be hard to distinguish this event from a straightforward increase in hormone B. In some instances a hormone sensitizes its own receptor to its own action (self-priming). This potentially explosive interaction ceases when a certain level is reached. As the amounts of stimulating hormone increase, sensitivity drops; the receptor is said to be 'down-regulated'.

There are some dozen hormones which are of particular importance in the control of growth; the actions of each will be described below. They are thyroxine from the thyroid gland; cortisol and adrenal androgens from the cortex of the adrenal gland; testosterone from the Leydig cells of the testis; oestrogens from the ovary; insulin from the islets of Langerhans of the pancreas; melatonin from the pineal gland; and a series of hormones from the pituitary gland. These are growth hormone (GH); thyroid-stimulating hormone (TSH), also called thyrotrophin; adrenocorticotrophin (ACTH); and the gonadotrophic hormones, which are follicle stimulating hormone (FSH) which stimulates growth of the young ovarian eggs and follicles of the ovary and the sperm-producing cells of the testis, and luteinizing hormone (LH) or Leydig-cell-stimulating hormone, which stimulates the growth and secretion of the corpus luteum in the ovary and the Leydig cells in the testis. The pituitary also produces prolactin, necessary for lactation, and vasopressin, which is necessary for control of body water. Another gonadotrophic hormone with actions similar but not identical to those of LH is produced by the cells of the chorion, which is the cell layer of the placenta nearest the maternal circulation. Chorionic gonadotrophin (CGT) enters primarily the maternal circulation, though it crosses from here into the foetus. The placenta also produces a hormone somewhat resembling growth hormone, called placental lactogen.

Recently it has become clear that many – perhaps most – of these hormones can exist in several slightly different forms. For example, FSH changes at puberty into a more biologically active form (or, what amounts to the same thing, a higher proportion of the more active form is secreted into the blood). Chemical estimates of hormone level (usually by radioimmunoassay) often fail to distinguish these forms, since they may use reactions against parts of the molecule different from the parts which have the biological action. Bioassays of hormones are expensive and laborious, but in the end unavoidable.

Many hormones, also, travel in the blood bound to carrier proteins. The amount of carrier may be altered by another hormone, which thus influences the first hormone. The degree to which dissociation occurs may also be important; usually it is only the free portion of hormone that is active.

Control of Endocrine Secretion

Most of the endocrine glands secret their hormones in response to messages, themselves hormonal, arriving from the pituitary gland. Thus the Leydig cells of the testis are stimulated to action by pituitary LH, the thyroid cells by pituitary TSH and so on. The pituitary, however (once known to generation upon generation of medical students as the 'conductor of the endocrine orchestra') is only the manufacturing site and storage depot of these stimulus hormones. Their control resides in the hypothalamus, a portion of the brain geographically close to the pituitary and connected to it by a special system of blood vessels (*see* Figure 32). These vessels are designed specifically to receive chemical messengers emanating from the tips of nerves in the hypothalamus and to pass them on to the cell receptors in the near-by pituitary. In the midst of this highly organized complexity it comes as a welcome surprise to learn that the hypothalamic nerve-secreted hormones are extraordinarily simple substances; so much so, in fact, that many can be synthesized on a commercial scale. The simplest of all is thyrotrophin-releasing hormone (TRH). This is a tripeptide, i.e. a molecule consisting of only 3 amino-acids. In contrast, human growth hormone (HGH), a fairly typical pituitary hormone from this point of view, consists of 191 amino-acids (though not, of course, all different ones). It seems a general rule that nerve cells can only synthesize and secrete rather simple substances, which serve as triggers to set off the larger, more complicated charges manufactured in the endocrine glands.

Each pituitary hormone awaits its own specific hypothalamic message; TSH is released by TRH, ACTH by CRF, corticotrophin-releasing hormone (or

Figure 32   Diagram showing position of pituitary gland immediately below, and joined to, the brain. The hypothalamus is the area of brain just above the pituitary. The circled inset shows the system of blood vessels going from the hypothalamus to the anterior part of the pituitary factor), growth hormone by GHRH. It seems that LHRH, luteinizing-hormone-releasing-hormone, a 10-amino-acid peptide, may be an exception, in that it released too also, though to smaller extent. For this reason, it is often called GNRH, gonadotrophin-releasing hormone. In addition two horm-ones, prolactin and growth hormone (which are of very similar structure and clearly derived from a common ancestral molecule in evolution), are controlled by being stopped instead of (prolactin) or as well as (growth hormone) being started. The inhibiting hormone for growth hormone is called somatostatin and consists of 14 amino-acids; GHRH has 40.

Releasing hormones cause both synthesis and release, normally, of the hormone they control, though the two actions seem to be biochemically distinct.

The hypothalamic releasing (or inhibiting) substances are themselves switched on or switched off in response to certain defined stimuli, of which the most important is the level of the relevant hormone circulating in the general blood-stream. This level is sensed by specialized cells in the hypothalamus near to the cells which synthesize and secret the releasers. Thus the feedback circuit is established: sensor → releaser → pituitary hormone → peripheral hormone → sensor. Outside influences can play on the circuit, particularly by altering the sensitivity of the sensor to circulating hormone. The way the events of puberty are controlled provides an example. At the beginning of puberty the sensor for oestrogen is set so that it responds to a rise in blood oestrogen, by turning off GNRH, when blood oestrogen is still at a very low level. Then impulses from other parts of the brain arrive which diminish the sensitivity of the sensor so that now all it responds to is a much higher level of oestrogen. Consequently, gonadotrophins and oestrogen rise until the new, pubertal, level of oestrogen is established.

There may be a number of sensors serving a single releaser. Furthermore, these sensors may change their relative importance at different times of life or in different circumstances. It seems, for instance, that in primates, just before menstruation occurs, the sensor serving FSH- and LH-releaser stops being turned off by oestrogen and is suddenly switched to being turned on by it (*see further* p.101 *below*). Hypothalamic sensors and releasers are still a rapidly developing area of knowledge, and there are certainly many sensors and probably some releasers of which we are still ignorant.

In certain circumstances some hormones are released or inhibited without the full circuit being involved. There is evidence that a rise in the level of blood thyroxine sometimes is sensed directly at pituitary level, thus switching of TSH immediately, without intervention of TRH.

Pulsatility of Hormone Secretion

Most hormones are secreted not continuously, but in episodic or pulsatile fashion, in bursts. Each burst typically lasts some 10-20 minutes. In between bursts the blood level is low, perhaps undetectable. Thus taking a single blood specimen is of very limited value in studying growth hormone, for example, or luteinizing hormone. Indwelling needles have to be put in place and samples taken every few minutes over periods of hours. Most peaks seem to be satisfactorily detected by samples taken every 10 or 15 minutes, and this is the time interval now usually used. (It raises problems of the total amount of blood withdrawn, particularly in children, and in consequence chemical methods are being continually refined to work on lower and lower blood volumes.)

For many hormones the frequency of the pulses is just as much part of the message to the activated organ as is the level of the pulse. Particularly is this so in the case of GnRH, LH and FSH (*see below*) as the elegant experiments of Ernst Knobil at Pittsburgh have shown. All the changes of puberty have been produced in infantile Rhesus monkeys by pulsed stimulation of the GnRH-secreting neurones, or by giving 'pulsed' injections of synthetic LHRH (one every 90 min). Continuous infusion of LHRH does not work; indeed continuous LHRH in the shape of a long-acting analogue given once daily is a recent highly succesful treatment for pathological precocious puberty; it turns off entirely the normal GnRH secretion and returns the child to the prepubertal state. The control of the menstrual cycle in adults depends on these serial alterations in hormone frequency. Thus for a thorough investigation a full 24-hour study needs to be done, with regular sampling. For some hormones however overnight sampling, from 9 pm to 9 am for example, gives almost as much information.

Growth Hormone (GH)

Growth hormone is necessary for normal growth from birth to adulthood. Children who lack it end up as adults about 130 cm tall, but with normal proportions; they have been called miniatures. Growth hormone, from 1958 to 1985, extracted from human pituitary glands and now made by recombinant DNA technology, can be given to such children by injection; they then grow at a normal rate, or a faster rate than normal if they are already small when the injections are started. (This is the phenomenon called catch-up growth, discussed further in Chapter 10.) Figure 33 shows the growth curve of a girl whose treatment started at 12·6 years. A further example is shown in Figure 88, p. 234.

Figure 33 Effects of growth-hormone deficiency on growth, with catch-up following treatment with human growth hormone (HGH) given over three periods indicated by arrows. Girl with isolated growth-hormone deficiency. Open circles represent bone age. F and M are father's and mother's height centiles, and bar represents predicted height centile range for their children. Puberty ratings at each age are shown by arrowheads, lower section

Growth hormone has the curious property that it is species-specific. Whereas insulin and ACTH from animals works perfectly well in humans, GH from animals does not. Thus before the advent of recombinant DNA technology or 'gene-splicing', human growth hormone had to be extracted at autopsy, since with 191 amino-acids it is too large a molecule to be synthesized chemically.

Growth hormone secretion characterises adults just as much as children, although the part it plays in adult physiology is a good deal less clear. In exercise, and perhaps more importantly from an evolutionary point of view, in starvation, it takes fat out of the fat depots and enables it to be metabolised to provide energy. Dr Ho and his colleagues at the University of Virginia fasted volunteers completely for 5 days, and showed that the amount of GH secreted per 24 hours went up to three times the normal level. In present-day culture adults with untreated GH deficiency, at least in the inherited form, seem to be perfectly healthy and complain only of being small. Nevertheless GH is amazingly abundant in the human pituitary, with levels 1000 times greater than those of any other pituitary hormone. Women secrete on average more GH than men from puberty to the menopause (the effect is oestrogen-mediated). The frequency of bursts however is similar in the two sexes (in Man; in rats it is very different). As both men and women get older their secretion rates diminish, chiefly, it seems, because of diminished stimulation by GHRH and perhaps increased inhibition by somatostatin. Here, as elsewhere, maturity seems to reside in the hypothalamus.

GH is a prime example of an episodically-secreted hormone. Most of the time the amounts of GH in the blood, of children as well as adults, are so low as to be scarcely detectable, but some 8 or 9 times in each 24 hours the levels rise sharply for the order of 10-20 minutes. The bursts come at irregular intervals, though there are usually more at night than during the day. The nearest thing to a regular burst is one which frequently appears shortly after entry to the deep phase of sleep.

Growth hormone secretion is increased by exercise, and also by anxiety. Its action in mobilizing fat from the fat depots and using it for producing energy, spares the carbohydrate stored in the muscles. It also causes growth of muscles, at least in high dosages, (wherefore it is sought after by body builders and unscrupulous athletes).

During puberty the amount of growth hormone secreted per day increases, reaching its highest level more or less coincidentally with the peak of the adolescent growth spurt, for which it is partly (in the case of girls, perhaps wholly) responsible. (The changes are all of amplitude not frequency.) After the end of the spurt the amount falls again, to the young adult level. The major stimulus to this increased secretion seems to be the rise in sex hormones, that is oestrogen in girls and testosterone in boys. The amount of GH secreted during 24 hours by perfectly normal pre-pubertal children has turned out to be surprisingly variable, at least from child to child (we know virtually nothing about from time to time). Some children, of average stature and growing at average velocity, seem to secrete levels of GH that were previously thought to be characteristic of partial deficiency. The endocrine problems of growth seem not, after all, to be entirely solved. At the time of writing it is unclear whether or not tall boys and girls habitually secrete more GH than average or shortish children: the studies so far reported are in disagreement. Nor is it clear, at present, whether giving excess GH by injection to normal healthy boys and girls with heights at the 3rd centile will turn them into healthy boys and girls with heights ultimately at the 50th (or even, more modestly, the 20th). Still less is it clear whether, if 'successful', such a treatment would be a good thing. The first writer of a textbook on human growth, Stoeller in 1729, dismissed such an idea with the curt comment that 'to tamper with God's work was impious, and what might not follow if everybody by artificial means could compel the growth of his body?' (*see* Tanner, 1981).

Somatomedin C – IGF1

Surprisingly perhaps, GH needs an intermediary in its action on the bones to make them grow. This intermediary is called somatomedin C (SMC) or IGF1 (for insulin-like growth factor, discovered simultaneously in Zurich while SMC was being identified in Sweden and the United States). Somatomedin is secreted by the cells of the proliferative zone of the cartilage, and growth hormone appears both to increase the number of these cells, and also to turn on their IGF secretion. (Another somatomedin, locally produced, is probably responsible for the action of growth hormone on muscle, and yet another for the action on skin.) Somatomedin C, which is also secreted by the liver, is a peptide of 70 amino-acids and has been manufactured by recombinant DNA technology. Its blood level is low in infancy, rises gradually till puberty, then increases more swiftly to reach a peak at peak height velocity, after which it falls to the (prepubertal) adult value. Thus the curve of blood level before puberty is exactly the opposite of the velocity curve. Perhaps this is because the active level is that found locally in the cartilage, with blood level as merely a spill-over product. In any case blood SMC also reflects things other than GH. It drops during fasting (whereas GH rises), and recovers when the subjects are once more fed, in particular if they are fed high-protein food. There is recent evidence, needing confirmation, that blood levels of somatomedin C are a little lower in short children than in tall ones (*see* Binoux *et al.*, 1986). It is certainly true that small poodles (toys) have much lower SMC values than tall poodles (standards). Perhaps we are getting near to a physiological explanation of individual differences in human stature.

Thyroid Hormone

Thyroxine, a small molecule available synthetically, is necessary for normal growth from early foetal life onwards, and for normal physiological function in children and adults. It begins to be secreted at about the 15th to 20th postmenstrual week and is essential in the foetus and very young child for protein synthesis in the brain and for the proper development of nerve cells. As the brain matures this action becomes less prominent. Thus children born with thyroid deficiency become mentally handicapped unless treated at once. Diagnosis from clinical appearance is difficult in the first crucial days after birth, but routine biochemical screening of all newborns is technically possible and has been initiated in many areas. The incidence of the disorder is probably about 1 in 5,000 births, which is similar to that of phenylketonuria (PKU), long screened with success. Treatment is very effective, so population screening is a major advance.

Children who develop a deficiency of thyroid hormone at a later age have increasingly slow bodily growth and delayed skeletal and dental maturity, but do not suffer any particular brain damage. On treatment with thyroxine, such children show a great catch-up in growth and eventually reach perfectly normal adult height provided they are treated with reasonable promptness (Figure 34). It does not seem that the intensity of normal growth is regulated by the level of thyroid hormone. The thyroid's action is termed 'permissive'; that is to say, a normal thyroxine level permits the cells of the body to function properly; when there is too little, most cells, including those of the pituitary, function too slowly to be effective.

The rate of secretion of thyroxine is regulated by pituitary TSH (thyroid-stimulating hormone) itself controlled by hypothalamic TRH. TSH is secreted in a pulsatile fashion, with around 6 bursts per 24 hours. Thyroxine level itself regulates TRH secretion, in a classical feedback loop.

Adrenal Hormones

The adrenal gland consists of an inner core, the medulla, and an outer shell, the cortex. The two are really quite separate glands from a functional point of view, and the medulla, which secretes mostly adrenalin, has little to do with growth.

The adrenal cortex secretes three groups of hormones:

1. Mineralocorticoids   These are aldosterone and to a smaller extent 11-deoxycorticosterone, and they maintain within acceptable limits the levels

Figure 34   Effect of thyroid hormone deficiency on height growth, with catch-up
following treatment with thyroxine. Puberty ratings not available
(Redrawn from Prader, Tanner and von Harnack, 1963)

of sodium and potassium in the tissue fluids. Aldosterone secretion is not
controlled by the pituitary, but by a local feedback, reacting to changes in
blood composition and volume. Aldosterone is essential to life, but
otherwise has no direct bearing on growth.

2. *Glucocorticoids* (or corticoids)   Cortisol (also called hydrocortisone) and, to a lesser extent, corticosterone, are hormones secreted in a pulsatile manner throughout life. In young adult subjects not under stress there seems to be eight to a dozen pulses during the daylight hours, and no pulses at night. Cortisol increases the formation of glucose from protein and has an anti-inflammatory and anti-stress action. The rate of secretion is raised to 2 or 3 times the ordinary level in response to the stresses of infection, extreme exercise or severe emotion. Secretion is controlled by ACTH from the pituitary, and ACTH is itself released by the action of corticotrophin-releasing factor from the hypothalamus. The administration of (synthetic) cortisol, or its near relative cortisone, to persons whose pituitaries have been destroyed is essential if they are to lead anything approaching normal lives.

Cortisol is secreted during foetal life and throughout childhood in much the same amounts as in adults, relative to body size. No particular increase occurs at puberty, apart from that expected from the child's increased size. However, cortisol does have an anti-growth action. Though in normal amounts it probably plays no part in controlling rate of growth, if given in excess of normal it slows up growth in height and retards skeletal maturity, while at the same time causing increase in fat. This may easily occur in children treated with corticoids for severe asthma, rheumatoid arthritis (Still's disease), kidney disease, or severe eczema. In all these diseases corticoids should be avoided if possible; but sometimes they are absolutely necessary. There is evidence that giving them every other day instead of every day has considerably less effect on growth and often suffices therapeutically. Using the Valk knemometer Wales and Milner, of the University of Sheffield, have actually demonstrated a consistent pattern of blocked growth on the steroid day, followed by increased growth on the following, steroid-free day. Cortisol appears to block the ability of growth hormone to produce somatomedin; and this cannot be overcome by giving excess growth hormone. Cortisol may also directly inhibit the action of somatomedin on cartilage cells and perhaps the secretion of growth hormone itself.

3. Androgens   These are three or four closely-related substances secreted by the adrenal whose actions are in most respects similar to those of testosterone, the male hormone. They are therefore called androgens ('andros' being Greek for man). Their function is not really known, despite the fact that they are secreted in considerably larger amounts than most other hormones. During early childhood the blood level of adrenal androgens (chiefly dehydroepiandrosterone and its sulphate) is low, but at about age seven the level increases in both girls and boys, not to a great degree, but nevertheless distinctly. This rise is sometimes referred to as adrenarche. What it signifies we do not know; perhaps it is related to the midgrowth spurt, indeed even as its cause. Or perhaps there is some connection with the fact that subcutaneous fat stops falling and begins to rise at about this age.

At puberty a second, and much greater increase takes place, which continues for rather longer than most of the other hormonal increases, of testosterone, for example. Thus adult values are reached at about 18 or 20 years. Then, beginning around age 40, the levels decline once more, until by 60 or 70 the prepubertal values are reached. There is, of course, much individual variability in when the peak is reached, and when the decline begins.

It seems likely that the adrenal androgens are partly responsible for controlling maturation of the skeleton, i.e. the advance of bone age. It used to be thought that they were the cause of the growth of pubic and axillary hair at puberty in girls, but more recent studies have cast doubt on this. Girls with precocious puberty, due to GnRH being secreted not at age 10 but at say 4, develop pubic hair as well as breasts (adrenal androgen remaining normal) and when the GnRH secretion is shut off (*see* p.89) the pubic hair stops growing and if it is only at stage 2 or 3 actually falls out (at the same time the breast regresses). It seems the sex steroids oestradiol and testosterone play a large part in controlling pubertal hair growth.

Many years ago it was shown that more muscular young men produced more androgens, probably including adrenal ones, than less muscular men, and it may be that the prime role of these hormones is to support the muscles.

The way in which adrenal androgen secretion is regulated is still something of a mystery. It is stimulated a little by ACTH but to a smaller extent than is cortisol secretion. The pubertal rise cannot simply be attributed to increase in ACTH, since cortisol does not rise concurrently. A special adrenal androgen-stimulating hormone has been postulated but never found.

Foetal Adrenal Cortex

During the foetal period there is a special zone of cells in the adrenal gland, lying between the interior medulla and the usual, exterior cortex. The zone appears at about 12 postmenstrual weeks, enlarges till birth or a little before, then swiftly atrophies, disappearing completely. At its maximum it occupies 80% of the whole adrenal cortex. It is dependent on and stimulated by a special pituitary hormone, only recently discovered and only present in the foetus, called CLIP, which stands for corticotropin-like intermediate-lobe peptide. It secretes exclusively dehydroepiandrosterone and its sulphate (i.e. the androgens of the adult adrenal). These substances go to the placenta, where they are converted to oestrogens, which seem to be necessary in considerably quantity for the normal development of pregnancy.

In this, as in some other interactions, foetus and placenta function as a co-operative unit, since each contains enzymes lacking in the other. Thus certain materials can only be formed by precursors passing through both sides of the unit. The foetal adrenal cortex is a specialization peculiar to primates and does not occur in other mammals.

Testosterone

Testosterone is the male sex hormone, secreted by the Leydig cells of the testis. There are three periods when its level in the blood rises. The first is in the foetus, where the rise begins at about 11 postmenstrual weeks and lasts probably till birth (*see* Chapter 4). During this time the testosterone first causes differentiation of the external genitalia to form a penis and scrotum, and then sustains the further growth of the penis. It may also have an action on the brain, differentiating the hypothalamus irreversibly to the male, non-cyclic, type (*see* p.55).

Blood testosterone falls rapidly in the first few days after birth but a second period of high blood levels occurs during the subsequent 6 months, the peak being reached about 2 months after birth. This rise has only recently been discovered and its function is still unknown. Indeed it may only be a rise in *total* blood testosterone, which includes testosterone being carried in the blood attached to a protein, and hence inactive. Free active testosterone may simply remain high throughout the perinatal period, coming down gradually to the 2-month mark. At any rate by 6 months of age the level is down again to just detectable amounts and it continues thus till puberty. Then a very large increase occurs. At this time testosterone is responsible for growth of the penis, prostate and seminal vesicles, the pubic, axillary and facial hair, the growth of muscles and, together with growth hormone, the adolescent spurt of the skeleton, especially the bones of the shoulders and vertebral column. Testosterone increases the output of growth hormone (just as oestrogen does, *see below*)

and also itself has an additional action on bones and muscle, this accounting
for the difference between the male and female adolescent spurts. Since
facial hair grows late and pubic hair early in the sequence, either the hair
follicles at the two sites are unequally sensitive to differences in dose
(pubic hair at low dose, facial hair at high dose) or else there is a sequential
maturation of testosterone receptors, this occuring earlier at the pubic site
than at the facial. The latter is the more likely possibility, since when
testosterone is given to a patient lacking it, pubic hair grows at once,
together with the penis and shoulders; but even at high doses the beard
responds only later. Testosterone secretion is controlled by pituitary LH
(luteinizing or Leydig-cell-stimulating hormone).

Oestrogens

The ovaries secrete female sex hormones, collectively known as oestrogens. The main one is oestradiol. Its blood level is low until puberty,
when a large increase occurs causing growth of the breasts, uterus and
vagina, and development of the associated vaginal glands. It also causes
growth of parts of the pelvis and of pubic hair. Its increase at puberty is also
responsible for the adolescent growth spurt in girls, in that it is the
oestrogen increase which is responsible for the increased output of growth
hormone at this time. Oetrogen may even have a direct action on
somatomedin c production, though this is uncertain at present.

When menstruation starts, oestradiol levels fluctuate regularly with the
phase of the cycle. Oestradiol secretion is controlled by pituitary FSH
(follicle-stimulating hormone).

Gonadotrophins

There are two pituitary gonadotrophins: follicle-stimulating hormone
(FSH) and luteinizing or Leydig-cell-stimulating hormone (LH). Both are
secreted at low levels during childhood and increase sharply at puberty.

In males, LH is secreted, like GH, in a pulsatile fashion. In bulls secretion
follows the sight of a receptive cow or even a perfumed model. It is
uncertain how far similar considerations apply to man. The first indications
of the pubertal rise are pulses of LH released during sleep; subsequently LH
is released during the daytime also, until by the end of puberty the total
amounts released by day and by night are about equal, with some 12 to 15
pulses per 24 hours. Blood testosterone rises after each LH pulse, with a
peak some 20 minutes later. Thus the first pubertal rise of testosterone
occurs in the early morning.

LH in females is secreted in a cyclic fashion and interacts with FSH to control the menstrual cycle. The egg grows under the influence of FSH and is shed; the cells immediately surrounding it remain behind in the uterus and form a body called the corpus luteum: LH stimulates this body to produce the hormone progesterone. Progesterone maintains the uterus in a state receptive to the implantation and growth of the ovum should it be fertilized; if no fertilization and implantation occurs, LH levels decline, the uterus sheds its lining in menstruation, and the cycle starts again. While the follicle is growing the frequency of LH pulses rises from about 15 per 24 hours to about 30 per 24 hours, but during the phase of the corpus luteum it declines progressively, finally reaching about 4 per 24 hours.

FSH in males causes growth of the seminiferous tubules, the sperm-producing parts of the testis. These occupy 90% of the volume of the testis, so the testicular enlargement at puberty is almost entirely due to them and not to the Leydig cells, however much the number of these increases. FSH is necessary for the growth of sperm, just as it is for the growth of eggs. Much of the initial growth of eggs takes place during late foetal life, and in girls, though not in boys, FSH is especially high around 20-30 post-menstrual weeks.

Both gonadotrophins are released from the pituitary by LHRH, hypothalamic luteinizing-hormone–releasing hormone, which also stimulates their synthesis and storage. The feedback which shuts off LHRH is through oestradiol in the female and through testosterone and probably yet another testicular hormone, inhibin, in the male.

Prolactin

Prolactin, secreted by the pituitary, is necessary for the secretion of breast milk in the adult, but seems to play little or no part in the endocrinology of childhood. The blood level in boys stays the same from early childhood onwards, with no change at puberty. In girls, the prepubertal level is the same as in boys, but a rise occurs in late puberty: perhaps this is responsible for the filling out of the breast beneath the protuberant aerola which causes the passage from the oestrogen-induced stage 4 to the adult stage 5.

Pineal Hormone

Melatonin and perhaps other substances are secreted by the pineal gland, a small organ folded into a cleft above the central portion of the brain. The mechanical-minded Descartes thought it was the seat of the soul; subsequent investigators thought it a functionless remnant of the third eye of ancient vertebrates. Perhaps Descartes has the last laugh, for it is now clear that the pineal is a major timing device, fitting the life of the animal to the seasonal changes of its ecological niche. It has a paramount role in controlling the successive 'puberties' of seasonal breeders.

In Man its role in growth is less certain. Melatonin is secreted much more at night than during the day, and this pattern seems to be established at about the end of the first year. As age advances the day–night difference declines, and some data shows the nocturnal surge diminishing specifically as puberty begins. It may be that melatonin or an allied hormone is the influence which restrains puberty from occuring for so long in Man, but this at present is a speculative view. Removal of the pineal in infant monkeys did not result in either an earlier or a later puberty.

Insulin
Insulin is produced by the islets of Langerhans, collections of cells in the pancreas. It causes glucose to be absorbed into cells, and stored as glycogen in the liver and muscles; lack of it results in sugar diabetes. It is present in children just as in adults and has no particular action as far as growth is concerned except that it must be present in normal amounts for normal growth to occur. The blood level, both fasting and after a meal, rises during puberty, coincidently with the height spurt, and falls back again when puberty ends. Diabetic children whose diabetes is well controlled by injected insulin and a suitable diet grow quite normally, but even a small degree of laxity in the control produces stunting and retardation of growth.

The Endocrinology of Puberty
The precise details of how all these hormones produce the changes of puberty are not fully known, though they have become much clearer in the last few years.

The adolescent growth spurt is caused, entirely in girls and partly in boys, by the increased secretion of growth hormone that follows the rise of oestrogen or testosterone levels. The sex-dimorphic components of the spurt – the male increase in shoulder width, greater increase in muscles, and so on – are due to testosterone itself. The growth of breasts follows rather precisely the level of oestrogens, and pubic hair also follows the level of sex steroids in both sexes. Testosterone, of course, is reponsible for the growth of the beard.

Initiation of Puberty

The first event in the sequence of puberty, immediately preceding the morphological changes, is an increased secretion of the two gonadotrophic hormones of the pituitary. This is caused by an increase in secretion of GnRH by the cells in or very close to the arcuate nucleus of the hypothalamus. At first, the GnRH produces a relatively small secretion of LH and FSH, either because the GnRH receptors in the pituitary cells need some stimulation to sensitize them or because the LH- and FSH-synthesizing cells require some practice. Gradually the response increases, the blood levels of LH and FSH rise, and the peripheral events follow.

The question, however, is what initiates the pubertal secretion of GnRH. The classical experiments of Harris and Jacobsohn in London in 1952 showed that it was the brain, and not the pituitary gland, which had to reach a certain stage of maturity before the sequence of pubertal events could be initiated. The pituitary of a new-born rat, when successfully grafted in place of an adult pituitary, at once began functioning in an adult manner; it did not have to wait until its own natural time of maturation had been reached.

At birth the GnRH nuerons in the arcuate nucleus (which can be identified by immunochemical staining) are clearly secreting, for LH and FSH appear in the blood in much the same pulsatile way as in adulthood. Sex steroid secretion follows the LH, FSH stimulation, and indeed in the male infant a little time after birth blood testosterone is almost at adult levels. Coincidently there is a small increase in testis size. The function of this neonatal testosterone peak is not known; it seems most likely that it is necessary for some sort of organisation of the brain to occur, but just what, we do not know.

Then, quite rapidly, all this activity dies away and from about one year in boys and two years in girls FSH and LH, testosterone and oestrogen, all fall to scarcely detectable levels. Practically speaking, GnRH has been turned off at source; very low-level and infrequent pulses, entrained to (or following) sleep may occur, but that is all. Things remain like this until the GnRH neurons are once more roused from their slumber, after an interval of some seven or eight years. Then, in the period before any morphological signs of puberty appear, GnRH secretion begins again in earnest, at first at low amplitude and only at night. The sensitivity of the cells secreting FSH and LH increases as time goes by, the amplitude and frequency of GnRH increases, and the nature of the FSH and perhaps also LH molecules changes a little so they become more potent. FSH and LH levels go up, FSH more than

LH at first, because low-frequency GnRH favours FSH over LH, high-frequency the reverse.

Once the GnRH arousal has started, however, a second controlling mechanism comes into play. This is a straight negative feedback, operated by the steroid hormones upon both pituitary gonadotrophin-secreting cells and the GnRH cells, but primarily the former. As GnRH causes FSH to rise, and FSH causes oestradiol to rise, the oestradiol chokes off the pituitary response, to keep the FSH secretion at a relatively low level. At first the cells which secrete FSH are rather easily inhibited in this way, but as puberty progresses their sensitivity gets less, they become, as it were, more and more thick-skinned. By the end of puberty it takes 6 to 15 times as much oestradiol to shut them off as it took to begin with. So FSH rises higher, and oestradiol rises higher until it is at a level which causes growth of the breasts, pubic hair and so on. What it is that causes this desensitization of the so-called 'gonadostat' we do not know.

Gradually the ovarian follicles grow larger under the increasing FSH stimulation, and gradually, before each dies back, the amount of oestradiol it secretes gets greater. Eventually one follicle grows larger than all the others, the oestrogen level rises and stays at an adult level and then something different happens. The negative feedback is flipped over into a positive one: instead of inhibiting the release of FSH and LH, the high level of oestrogen causes a surge of LH and ovulation occurs. The positive feedback occurs only late in puberty, at menarche or a little after. Some 60–90% of menstrual cycles in the first year after menarche are anovulatory, that is, take place without an ovum actually being shed.

In children who lack the ability to secrete GnRH, or are excessively delayed in beginning their puberty, all these changes can be brought about by pulsatile injection of synthetic GnRH, as Charles Brook and his colleagues at the Middlesex Hospital in London, have shown (Stanhope *et al.*, 1987-1988). The treatment begins with the pulses every 90 minutes at night only, (twice as large in girls as in boys) and about midway through puberty (B3 in girls; testis 8 ml in boys) changes to pulses throughout the 24 hours, with the amount per pulse also increased. This way a more rapid than normal transit through puberty is obtained (about 1·5 years), but the sequence of events follows the usual course.

What it is that turns off the GnRH-secreting cells, and what turns them on again, we do not know. Whatever it is, it has played a large part in human evolution; as we go from monkeys to apes to man the interval between birth and puberty has progressively lengthened, providing for a longer and longer period of learning before reproductive life begins.

During the turned-off period some sort of timer, some tally of the degree of maturation, continues to tick away. Puberty is not something which appears suddenly out of a clear blue sky; it is the culmination of a long train of timed events. One thinks of a series of clocks, each linked to the next, which have successively to run their courses before their strikers fall. Chronological age is a poor guide to the state of advancement of the clocks, as has been explained in the previous chapter. Bone age is a better guide, at least as far as menarche is concerned (*see* p.79). Indeed Drs W. A. Marshall and Y. de Limongi (1976) of the Institute of Child Health have shown how bone age may be used in conjunction with chronological age to predict when menarche will occur. Chronological age enters the prediction because the older the girl, the more likely is menarche to occur soon. Bone age aside, British girls less than 13·0 years old are most likely to have menarche at 13·0, but those more than 13·0 years old are most likely to have menarche tomorrow. This is a somewhat unusual situation in predictive statistical work and failure to recognize that this is so has led to various invalid claims, for instance that age at menarche can be predicted from height and weight or that menarche takes place at a certain 'critical' value of weight, or weight for height (sometimes erroneously referred to as 'fatness'). Bone age is the only factor we know which does link more closely than chronological age with age at menarche, yet it itself is clearly only a symptom of some underlying maturity state, presumably located in the brain.

The nature and control of this maturity state are both obscure. One feels we should be able to specify a hormonal cause of early and late maturation, and isolate and synthesize a tempo-hormone. Some paediatric endocrinologists think that tempo is controlled by the total amount of growth hormone secreted, just as others believe that size is controlled this way. (Both cannot be right.) But really we are far from understanding the physiology of maturation. Children differ in how closely linked all their pubertal events are; in some the 'clocks' for breast growth and for menstruation run far apart, in others they are closely synchronized. External events may slow down or stop the clock. The stimulus which changes sensitivity of the oestrogen sensor comes from the nervous system. Other nervous impulses, sometimes contingent on stress and emotion, may interfere with the smooth running of events. Thus R. H. Whitehouse and the writer have seen quite unusually 'fractured' curves of growth and pubertal development in girls translated to unfamiliar boarding schools at various times in puberty.

CHAPTER 8

Growth and Development of the Brain

Despite the overwhelming importance of the brain in the life of man we know all too little about its growth and about the development of its intricate organization. Anatomical studies of brain structure are immensely laborious and few workers have had the courage, persistence and technical support needed to carry out thorough morphological studies of the brains of children or even, indeed, of young animals. Though biochemical studies of brain composition are much easier they inevitably gloss over just that detail which is the chief feature of the brain's organization. The time is certainly coming when immunochemical methods will allow the same sort of fine distinction between individual cells that the microscope and the tiny probes of the electrophysiologist permit. But at present structural studies remain the solid base on which understanding of the brain rests. This means making three-dimensional reconstructions based on light- and electron-microscopy, aided by various computer methods. The elegance and beauty of the pictures thus produced, for example by Sidman and Rakic (1974) of Harvard University, make molecular biology look dull and molecular structure simple.

There is great difficulty in obtaining a sufficient number of brains to make a study of normal development from children dying of accidental causes, or at least of diseases which do not affect the brain. Foetuses obtained from abortions are relatively numerous and the early intra-uterine portion of the curve of brain growth is accurately known. But the important part of the curve in the last weeks before birth, and in the period from birth to 2 or 3 years, is still poorly mapped and its details are in some dispute.

Growth of the brain is reflected, albeit somewhat indirectly, in growth of the skull, and for this, at least, we have accurate postnatal data. We also have good prenatal longitudinal data from measurements by ultrasound (see Chapter 3, p.47) of maximum head breadth and of head circumference. The head develops earlier than any other part of the foetal skeleton, and the peak velocity of head breadth is probably reached a little before 13 postmenstrual weeks, though a relatively high velocity continues till about 30 weeks. Velocity curves for length and breadth of the cerebrum also show maxima at about 13 weeks, which is to be expected since experimental work has shown that the growth of the vault of the skull depends on, and is controlled by, the growth of the brain below it.

Head circumference has its peak velocity at about 15 to 17 post-menstrual weeks, a little later than head breadth. This may be because the cerebellum, situated at the back of the skull (and hence reflected in head circumference though not in head breadth measurements) grows later than the cerebrum. The relatively high velocity continues till 32 to 34 postmenstrual weeks, but after this the deceleration is very rapid. Six months after birth, head circumference velocity is down to about 15% of its 34-week value, and by 1 year after birth, to 7%.

Head volume, representing brain size, calculated from a series of diameters and/or head circumference, has an s-shaped curve of growth from 12 weeks to 40 weeks, with a peak velocity located at about 30 weeks, and a fairly rapid decrease thereafter.

The velocity curve for the weight of the whole brain parallels that of head volume and reaches its peak around 32 postmenstrual weeks, according to the most numerous data, summarized by Dr Donald Cheek of Melbourne (see also Schultz et al., 1962). Its deceleration thereafter is quite swift; by 1 0 years the velocity is reduced to about 25% of the peak value and by 2 0 years to less than 10% of the peak. The early development of the brain compared with that of most other organs means that from early foetal life its weight is nearer to its adult value than any other organ except perhaps the eye. At birth the brain averages about 25% of its adult weight; at 6 months nearly 50%; at 2 years about 75%; at 5 years 90%; and at 10 years 95%. This contrasts with the weight of the whole body, which at birth is about 5% of the young adult weight, and at 10 years about 50%.

However, total brain size or brain weight is not a very satisfactory measurement. Different parts of the brain grow at different rates and reach their maximum velocities at different times. There are maturity gradients in the brain no less than in the bones of the skeleton. Figure 35 shows the percentages of total brain volume contributed by the pons and medulla (the hind part of the brain), the mid-brain, the cerebrum (which here includes corpus callosum, basal ganglia and diencephalon with the thalamus and hypothalamus) and cerebellum from three months of foetal life to adulthood. The hind-brain and mid-brain are most advanced, and the cerebrum next (for a picture of the brain, see Figure 36). The cerebellum has its peak velocity considerably later than the cerebrum, an arrangement which seems to be general among mammals. (The cerebellum is the part of the brain concerned chiefly with the control of fine movements; it has a more uniform structure than the cerebrum and seems to many investigators a somewhat easier proposition to tackle.)

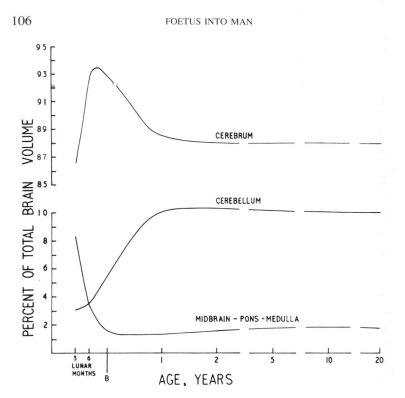

Figure 3.5   Percentage of total brain volume contributed by parts of the brain at ages from
3 postmenstrual months to adulthood. Cerebrum includes hemispheres, corpus striatum and
diencephalon
(From Tanner, 1977; data drawn from Dunn, 1921)

Even these curves are still far too non-specific. The brain is made up
of two fundamentally different sorts of cells, the neurons proper (or nerve
cells), and the neuroglia (or supporting cells). The neurons are the cells
which transmit impulses and synthesize and transmit neuropeptide
hormones. Each consists of a cell body, with a nucleus as in other cells,
and cytoplasm drawn out into a very large number of fine wire-like
processes called dendrites. In most cells one such process is larger and
much longer than the others and is called the axon. In a motor nerve of the
cerebral cortex, for example, the axon runs all the way down from the head
to cells in the spinal cord. Axons themselves usually have many smaller
branches. The dendrites are mostly short and form a pattern of arborization
or branching, the intricacy of which invokes comparison with the statistics
of the solar system. There are about 10^{12} or a million million neurons in
the brain, and the average cerebral cortex neuron would seem to have

some 30,000 nerve processes terminating on it, probably coming from one-tenth or so of that number of cells. The neurons are not tightly packed together like cells in most other tissues; there are narrow gaps between each of them, filled with tissue fluid. The connection with other cells is made through branches of the axon coming into close proximity with the dendrites of other cells, though without joining. Messages are passed across the tiny gaps by chemicals released at the nerve endings. It is thought that the most important clue to the functional capacity of a brain is the 'connectivity' of its neurons, i.e. the number of connections each cell makes with the others.

Recently it has been shown, by Dr A. Breuer of the Cleveland Clinic and others, that the axons of peripheral nerves carry a second set of messages, over and above the nervous impulses. Within each axon are separate parallel tracks, varying in number from 2 to perhaps 15, and along these tracks travel 'organelles', that is visually defined particles probably composed of protein-like substances. Some tracks are 'down', from cell body to neuro-muscular or neuro-sensory junction, and some are 'up', returning material to the nerve cell. On occasion a track may be used in both directions. The organelles travel quite slowly, on the order of 5–10 mm/hr. Precisely what they do is unknown, but the continuous traffic is clearly essential for the integrity of the muscle and the nerve cell. It is not yet known whether similar traffic occurs in the axones of the central nervous system, but the assumption must be that it does.

The neuroglia, which occupy about half the cellular volume of the brain, do not carry messages like the neurons. They are the support links, in the logistical sense, and act as intermediaries between neurons and the blood supply; a number have one process attached to a capillary blood vessel and the other intertwining amongst the surrounding neurons. They are smaller and more numerous that the neurons. They probably transmit glucose, amino-acids and other substances into neurons for the production of energy and the manufacture of structural protein and the chemical messengers. There is also evidence that during development some sorts of neuroglia act as scaffolding along which neurons move from their point of production to their final position. Other neuroglia manufacture the myelin sheaths which surround the axons. Thus neuroglia have clearly a crucial role in the nervous system.

Neurons and neuroglia develop at quite different times, at least as concerns cell division. Most of the neurons in the cerebrum are formed during the period from about 10 to 18 postmenstrual weeks; that is to say, their nuclei are formed then, surrounded by minimal cytoplasm. Axon and dendrite growth comes later; in some cases, it would seem, very much later indeed. Glial cells in the cerebrum begin to be formed around 15 postmenstrual weeks and continue being formed, though at a decreasing rate, until some time in postnatal life. It is said that new neuroglia cells are formed in the cerebrum up to about 2 years postnatally, but, surprisingly, in the cerebellum only up to about 15 months. However, these are generalizations based on chemical rather than structural evidence.

Recently Laurence Becker and his colleagues in Toronto made a quantitative study of the growth of the dendrites of neurones in two cell layers   (III and V) of the visual   cortex in children, from 20 weeks gestational age to seven years. All developed the typical branching pattern. The dendrites in layer V were always ahead of those in layer III; at 20 weeks there was no branching in layer III but already a little in layer V. In both layers the greatest increase of development, both of length and number of branches was between 36 and 40 weeks. No new orders of branching were added after birth in dendrites coming from the base of the neurones in either layer, but an average of three new orders were added to dendrites coming from the apex of these (pyramidal) cells. Thirty-five per cent of the total dendritic length was reached at birth in layer III, 55% in layer V. The childhood total was reached at four months in layer V, but not until two years in layer III. Similar studies by Peter Huttenlocher of Harvard University, showed that the total volume of the visual (striate) cortex was about 0·5 cm at 28 weeks foetal life, 1·7 cm at birth, 2·3 cm at 10 days, 3·0 cm at 2 months and 6·5 cm at 4 months. Thereafter there was no further increase (this being one of the 'earliest' parts of the cerebral cortex). The number of synapses per cm doubled between 28 weeks and birth; was static till two months and then nearly doubled again between 2 and 4 months. It reached a peak at 8 months, and thereafter declined: between this age and 11 years (after which no further change occurs) about 40% of the synapses are eliminated. These morphological data, he says, agree well with the behavioural data indicating that some plasticity of the visual cortex is retained up to about age 6.

Cerebral Cortex Development

We owe much of our knowledge about the development of the struc-ture of the cerebral cortex to the enormously painstaking studies of J. L. Conel, of Boston, who between 1939 and 1967 published analyses of the cortex at birth, 1 month, 3 months, 6 months, 15 months, 2 years,

4 years and 6 years. Additional studies of pre-term babies of 5, 6, 7 and 8 months, and of children of 8 to 22 years, have been added by Conel's associate, T. Rabinowicz.

The cortex is identifiable at about 8 postmenstrual weeks, and by 26 weeks most of it shows the typical structure of six somewhat indeterminate layers of nerve cells (the grey matter) on the outside, with a layer of nerve fibres (the white matter, consisting of outgoing and incoming axons) on the inside. At first the cells are small, consisting mostly of nuclei, with few and small processes. As they develop, axons and dendrites grow, protein appears in the cytoplasm, and many of the axons acquire an insulating sheath of myelin, the thicker fibres in general getting more, the thinner ones less.

From these cellular changes a series of criteria for maturation of parts of the cortex can be developed in just the same way as the criteria for skeletal maturity are developed from a consideration of the appearance of the bones. Conel used nine such criteria, amongst them the 'density' of neurons (i.e. the number of neurons per unit volume of cortex), which decreases as axons and dendrites grow in between them; size of neurons, which increases; nature of substances in the cytoplasm; length of axons and dendrites; and degree of myelination.

Two clear gradients of development occur during the first 2 years after birth. The first concerns the order in which general functional areas of the brain develop, the second the order in which bodily localizations advance within the areas. The most advanced part of the cortex is the primary motor area located in the pre-central gyrus (*see* Figure 36). This is the area whose cells initiate most movement. Next comes the primary sensory area in the post-central gyrus where nerve fibres mediating the sensation of touch end; then the primary visual area in the occipital lobe where nerve paths from the retina end; then the primary auditory area in the temporal lobe. All the association areas (where the primary impulses are compared and integrated with other impulses) lag behind the corresponding primary ones. Gradually the waves of development spread out, as it were, from the primary areas. Thus in the frontal lobe the parts immediately in front of the motor cortex develop next and the tip of the lobe develops last. The gyri on the medial surface of the hemisphere and in the insula are generally last to develop; the hippocampal and cingulate gyri lag behind most areas of the frontal lobe; and the insula develops later than any part of the frontal lobe examined.

Within the motor area, nerve cells controlling movements of the arms and upper trunk develop ahead of those controlling the leg. The same is

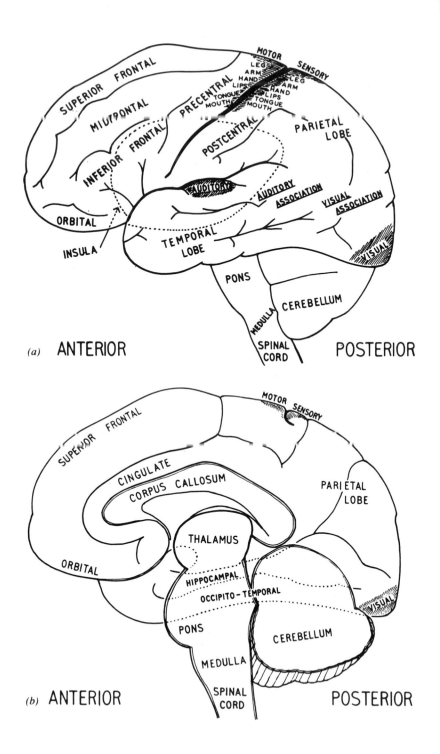

(a) ANTERIOR

POSTERIOR

(b) ANTERIOR

POSTERIOR

true in the sensory area. This corresponds to the greater maturity of the arm relative to the leg in bodily development and also to the infant's capacity to control his arms more than his legs. The leg areas remain the least developed up to 2 years and presumably somewhat beyond. In the association areas we would not expect gradients of this sort, since little or no localization by bodily areas occurs there. Conel, in fact, describes development outside the primary areas as proceeding in a fairly uniform sequence within each lobe, so that the cortex at birth is little developed and its appearance does not suggest that much cortical function is possible. By 1 month the appearance of the cells of the primary motor area of the upper limb and trunk suggests it may be functioning. By 3 months all the primary areas are relatively mature. At this age the motor area is the most advanced part of the cortex, and within it the hand, arm and upper trunk are the most advanced sections. The precise timings should be looked on as approximate, pending further research.

D. P. Purpura, also of Harvard University, found that the cells of the primary visual area of the cortex had their chief burst of maturation in the relatively brief period between 28 and 32 postmenstrual weeks. At this time the dendrites grew and 'spines' appeared on them, which is believed to signify full maturation; thus the visual cortices of the pre-term 8-month-old and the full-term neonate are said to be closely similar. He places development of the motor cortex also a little earlier then Conel, without disputing, however, the maturity gradients within the cortical areas.

To return to Conel's series; by 6 months most areas appear more mature than at 3 months and many of the exogenous fibres coming to the cortex have become myelinated, i.e. covered with an insulating sheath which in many nerves is essential for proper conduction of impulses. Between 6 and 15 months the rate of development is greatest in the temporal lobe and in the cingulate gyrus and insula; next greatest in the occipital lobe; and least in the parietal and frontal lobes, which have already passed through most of their development. The primary motor area is still slightly in advance of all the others, but within it the leg still remains behind the rest. The child's behaviour reflects this; his is beginning to control his hands and arms quite successfully, but he controls his legs less well. The visual

---

Figure 36 (*Opposite page*) Lateral and medial views of the brain, to show divisions of cerebral cortex, areas of localization of function, and the location of thalamus, cerebellum, pons, medulla and spinal cord: (*a*) lateral view of cortex; (*b*) medial view of cortex and subcortical structures
(From Tanner, 1977; after various maps)

association area has matured somewhat and is ahead of the auditory. By 2 years the primary sensory area has caught up with the primary motor, and the association areas have developed further. But some areas, most notably the hippocampal and cingulate gyri, particularly the insula, are still quite immature.

It is clear from the studies on myelination by P. I. Yakovlev and his colleagues that the brain goes on developing in the same sequential fashion at least till adolescence and perhaps into adult life. Myelination of nerve fibres is only one sign of maturity, and fibres can and perhaps sometimes do conduct impulses before they are myelinated. But where the two studies overlap the information from myelin studies agrees well with Conel's information on nerve cell appearances. As a rule the fibres carrying impulses to specific cortical areas myelinate at the same time as those carrying impulses away from these areas to the periphery; thus maturation occurs in arcs or functional units rather than in geographical areas.

A number of tracts have not completed their myelination even 3 or 4 years after birth. The fibres which link the cerebellum to the cerebral cortex and which are necessary to the fine control of voluntary movement begin to myelinate only after birth, and do not have their full complement of myelin till about age 4. The reticular formation, a part of the brain especially developed in primates and man and concerned with the maintenance of attention and consciousness, continues to myelinate at least until puberty, perhaps beyond. Myelination is similarly prolonged in parts of the forebrain near the mid-line. Yakovlev suggests that this is related to the protracted development of behavioural patterns concerned with metabolic and hormonal activities during reproductive life.

Throughout brain growth from early foetal life the appearance of function is closely related to maturation in structure. Fibres of the sound-receiving system (the 'acoustic analyser') begin to myelinate as early as the 6th foetal month, but they complete the process very gradually, continuing until the 4th year. In contrast, the fibres of the light-receiving system or optic analyser begin to myelinate only just before birth but then complete the process very rapidly. Yakovlev points out that in foetal life the sounds of the functioning of maternal viscera are the chief sensory stimuli. They are evidently not perceived at a cortical level; but at a subcortical one the analyser is working. After birth, however, visual stimuli rapidly come to predominate, for man is primarily a visual animal. These signals are very soon admitted to the cortex; the cortical end of the optic analyser myelinates in the first few months after birth. The cortical end of the acoustic analyser, on the other hand, myelinates slowly, in a tempo probably linked with the development of language.

There is clearly no reason to suppose that the link between maturation of structure and appearance of function suddenly ceases at age 6 or 10 or 13. On the contrary, there is every reason to think that the higher intellectual abilities also appear only when maturation of certain structures or cell assemblies, widespread in location throughout the cortex, is complete. Dendrites, even millions of them, occupy little space, and very considerable increases in connectivity could occur within the limits of a total brain weight increase of a few per cent. The stages of mental functioning described by Piaget and others have many of the characteristics of developing brain or body structures and the emergence of one stage after another is probably dependent on (i.e. limited by) progressive maturation and organization of the cortex.

The introduction of the technique called positron emission tomography (PET) has recently made more functional mapping of brain growth possible. A positron-emitting isotope, usually labelled 2 deoxy-fluoro-glucose, is injected, and the intensity of positron emission measures the glucose utilization in discrete parts of the brain. Harry Chugani and his colleagues in Los Angeles showed that during the first month after birth the most metabolically active parts of the brain were the sensorimotor cortex, the thalamus, the brainstem and the deeper part of the cerebellum. By three months, the anterior parietal, calcarine (primary visual) and temporal cortices, with the basal ganglia and cerebellar cortex were becoming active, and by six to eight months the frontal cortex and the secondary visual parts of the occipital cortex. The metabolic rate of the grey matter regions was low at birth but then rose rapidly, reaching and passing the adult value by 2 years. The peak was reached at about age 9, when a decline to adult values began. The time course matches that to be expected on the basis of the over-production and subsequent elimination of neurones, synapses and dendritic spines in the central nervous system.

Hemispheric Specialization

The two hemispheres of the human brain are not mirror images of each other. Some tasks are predominantly done by one side, some predominantly by the other. The brain itself is structurally asymmetrical to a far greater degree than the limbs or eyes or ears. The upper surface of the temporal lobe, as also the whole occipital region, is on average considerably larger on the left side than on the right, a difference established during foetal life. The temporal lobe area is that which is chiefly concerned with the analysis of sounds, and indeed it is the left hemisphere which predominates in the activities of receiving, processing and producing, language. As Levy puts it 'The right hemisphere may know that "cat" means a furry small pet with claws, but it does not know that "cat" rhymes with "rat".' That association is made in the left hemisphere. In contrast, the right hemisphere plays the major part in the processing of spatial information, either visual or tactile. It notes visual similarities in preference to conceptual ones; it codes input in terms of images while the left hemisphere codes input in terms of linguistic descriptions. The relation of brain differentiation to handedness is complex and not properly understood at present, but some psychologists working in this field think that the brains of lefthanders are less clearly lateralized than those of righthanders, and most are agreed that severe problems of school abilities and behaviour may arise if differentiation of the hemispheres fails to take place normally. S. F. Witelson, for example, thinks that developmental dyslexia is due to having 'two right hemispheres and none left'.

There are several instances in other organs of the body where lateralization is dissimilar in the two sexes, and there is some evidence that there are more women than men with larger right than left occipital regions (the atypical, or minority pattern). Witelson has found evidence of a sex difference in the timing at which lateralization of spatial information processing is established. Boys from at least age 6 (the lowest age tested) processed tactile spatial information better in the right hemisphere; girls processed it equally well in both sides at least till age 15 (the highest age tested). The average of the two sides was the same for both sexes. This work needs confirmation and extension, but may be of considerable importance in demonstrating a prepubertal sex difference in brain function, in the sense of one sex specializing earlier than the other. Boys have long been known to have a higher incidence than girls of reading and language disabilities, and reading, at least, involves spatial as well as linguistic processing. It would be quite in keeping with other aspects of sex differences if the male specialized away from the basic pattern, and ran some excess risk in doing do.

Influences on Brain Development

To what extent environmental stimulation can influence brain maturation or organization is not entirely clear. Ramon y Cajal, the great Spanish pioneer of brain histology, believed simply that use of a cell caused axons and dendrites to grow, or, in modern terms, increased a cell's connectivity. Evidence in favour of his view is gradually accumulating. Pysh and his colleagues have demonstrated that exercising young growing rats resulted in a larger dendritic tree of the cells in the cerebellum subserving movement that that seen in more sedentary rats. Indeed, a body of knowledge has emerged which indicates, perhaps, the beginning of a true understanding of how the nervous system generates its fantastically intricate and precise patterns.

Recently increased attention has been paid to the fine analysis of brain development in animals carrying genes causing brain abnormalities. As usual, the study of abnormality throws much light on normal processes; what happens in the absence of a certain class of cells, for instance, may give clues as to what that class of cells normally does (and this is the elementary level we are still at in studies of the brain).

The picture is one of much more movement and change within the central nervous system, or at least within the growing central nervous system, than we had previously imagined. The fact that no new nerve cells can be made after the first few months of life had somehow produced a picture of immutability wholly unlike our picture of, for example, liver cells or cells of the thyroid gland. Yet the dynamic state of the body constituents holds true as much for the brain as for any other part of the body. The turnover of amino-acids in even the adult brain is as high as that in the liver, and higher than that of any other tissue of the body. Since to supply its energy the brain only uses glucose, this means that the protein in the cell is continuously being synthesized and degraded. (A small amount of amino-acid is made into transmitter substances but not nearly enough to account for the massive turnover.) Probably part of this is to generate the organelles that presumably travel down the spinal cord axones, and to degrade the organelles returned. The turnover may also indicate that overt structural changes are occurring, of the sort that are in principle recognizable by the electron microscope. It is now clear that synapses, at least in certain areas of the brain, can be shed like autumn leaves, to bud again at the call of a hormone (*see below* p.118).

Much of the evidence has come from studies of the visual system. In all mammals very precise connections are made from the retina to the occipital part of the cerebral cortex. In those mammals such as the monkey and man, whose eyes are situated in the front rather than the side of the head so that the visual fields overlap and binocular vision is possible, half the nerve fibres from each retina cross to the opposite side in travelling to the cortex. This ensures that each tiny section of the cerebral cortex has fibres from both eyes terminating in it. Thus an object in space which is seen by both eyes generates impulses in each retina and these two sets of impulses end up exciting precisely contiguous sets of nerve cells in an area of the cortex. The visual cortex, indeed, consists of millions of columns or slabs of cells arranged with crystalline regularity, each column being a cellular machine equipped to analyse the visual events that occur in a tiny portion of each retina. The location of the event is given by the identity of the particular slab activated and the neurons in the slab analyse what type of event is occuring. The problem is to discover how this extraordinary regularity has been generated during growth.

An equal regularity is seen in a portion of the brain called the geniculate body, which is simply a relay station for nerves on their way from retina to cortex. The first set of fibres terminates there, and the impulses are taken up by a second set of neurons, which pass them on to the visual cortex. In the rhesus monkey, the cells of the geniculate body are all generated within the space of a few days during early foetal life, and Rakic (1976) has shown that at first they are not separated into their distinct layers. The fibres from the left and right retinae arrive, and, to begin with, all is confusion, fibres from one side being diffusely intermingled with fibres from the other. Gradually the ordered chess-board pattern emerges, with clusters of fibres from one side alternating with clusters from the other. Experiments in other species have led to the belief that two sets of endings terminating on one neuron must somehow exchange information as to their precise place of origin. On this information depends the retention or loss of the endings. It is not too hard to imagine how this might happen in an animal subjected to visual experience during this time; if two terminations were often activated simultaneously because they represented corresponding parts of the left and right retinae, they would persist; if not, one would die or at least become inactivated. It is harder to understand how this could occur in the darkened world of the embryo. D. H. Hubel and T. N. Wiesel, at Harvard University, have shown that in the monkey the whole elaborate structure of the visual system is complete before birth. However, covering one eye in the visual cortex from birth onwards results in a diminution of the number of cell terminations subserving the covered side, and an enlargement of the clusters of terminations from the other side. If the eye is covered for a year no recovery is possible. It seems that nerve cells, like muscles, atrophy from disuse.

In the cat, much of the architecture is complete before the eyes are opened, but C. Blakemore and others at Cambridge University have demonstrated that the cell connections required for precise binocularity or stereopsis of vision need the stimulus of visual experience. They also have shown that a kitten brought up in a world consisting exclusively of vertical stripes develops many more cortical cells which analyse this dimension of shape than cells which analyse the horizontal dimension (whereas normally each are present equally). This can be generalized. Simons and Land (1987) showed that trimming the whiskers of a rat from birth onwards resulted in pronounced abnormalities of the adult sensory cerebral cortex: again an expected environmental input cannot with impunity be forgone.

Furthermore, it seems that during development some axons form transient connections with a whole series of cells before establishing their stable adult contacts. In toads, for example, the eyes move their positions relative to the head during growth, and in order to keep the images seen by the two eyes as they grow projecting onto the same area of the brain, a series of cell contacts are successively made and broken. In addition there is evidence in adult mammals that axons entering the spinal cord from the periphery make a much wider series of connections in the spinal cord than are illustrated in the anatomical textbooks. These terminations seem normally to be suppressed; they remain, in Merrill and Wall's phrase, as 'ghosts of the cell's childhood'. The conditions under which they may be activated again are at present not clear.

R. Mark, of the Australian National University, whose book *Memory and Nerve Cell Connections* is a masterly exposition of the subject, sums up the new view thus: 'The principle of supplying, by growth, an overabundance of nerve cells or synaptic connections and fixing the final patterns by discarding many of them seems to show up often in the development of the nervous system, especially in the fine details of connection formation'. A recent study of the song control nuclei in the bird brain illustrates this point. In zebra finches only the males sing, and the location of the nuclei of cells which subserve this function is well known. Konishi and Akutagawa (1985) counted the number of cells in one of these nuclei (the nucleus robustus) in recently hatched and in adult birds. At day 15 males averaged 11,000 cells and in the adult this number had increased to 14,000. In the female however, the 10,000 present at day 15 had decreased to 3,000 in the adult, and these 3,000 were smaller than before; most of the cells had died or were dying. The growth in the male occurred just when the nucleus became innervated by axons from another nucleus in the brain, an innervation that did not occur in the female.

In the male canary two similar song control nuclei nearly double in size in the spring, when the birds produce their stable adult song. Each autumn, at the time of the moult, the cells diminish in number and size. The canary learns a new song each year, by improvisation and imitation of other males. Dr Nottebohm (1981) thinks that it is the shedding and regrowth of the synapses which permits this to happen: he compares it to the autumn leaf-fall of trees.

A similar cell death occurs in the spinal cord cells of the rat which innervate the bulbocavernosus muscle, a muscle used only by the male, in copulation. There are about 200 cells in the adult male nucleus, as opposed to 60 smaller cells in the adult female. It is testosterone secreted neonatally which causes the male cells to multiply and grow; and even in the adult this receptivity to hormone persists. Castration decreased both the size of the cell body and the length of the dendrites; both were restored by testosterone (Kurz *et al.*, 1986).

As for pathological influences, particularly careful studies of the effects of various agents on the rat's cerebellum have been made by Altman and his colleagues in Indiana and by Balazs in London (*see* Patel *et al.*, 1975). The rat is born so much earlier in its development than man that the cell multiplication phase of most of the cerebellar neurons as well as that of the cerebellar glia takes place after birth. Experimental manipulations are therefore relatively easy to make. A very large dose of radiation delivered daily to the cerebellum from birth to 10 days post-partum practically stopped cell multiplication, and the weight of the cerebellum at the end of the treatment was reduced to 47% of normal. However, catch-up then occurred, both in weight and in the number of cells formed. The catch-up was not complete; weight and cell number were both about 60% of normal at 23 days. It seemed that the duration for which cell division could continue was not extended; the cell division clock was adhered to, and the animal ran out of time in which to make cells. This is a result consistent with studies on the effect of undernutrition in rats. However, in another experiment, Balazs himself found the cell division period of the cerebellum to be stretched out somewhat by undernutrition, so the matter remains in doubt. The radiation had little effect on cells already present when it started; they continued to increase in size quite normally.

CHAPTER 9

The Interaction of Heredity and Environment in the Control of Growth

Growth is a product of the continuous and complex interaction of heredity and environment. Modern biology has no use for simplistic notions, let alone the obscurantist slogans of the deliberately ignorant. Statements such as 'Height is an inherited characteristic' or 'Intelligence is the product of social forces' (or vice versa of course) are intellectual rubbish, to be consigned to the trash-can of propaganda. As explained in Chapter 2, what is inherited is DNA. Everything else is developed. We may write $G * E \xrightarrow[\text{time}]{} (GE)$ where G stands for genetic and E for environmental factors. The product (GE) develops through time, and the way G and E are related is shown by a star, not a plus sign nor even a multiplication. We say that the interaction of genetics and environment is non-linear, meaning that the effects do not in general sum (as would be indicated by a plus sign).

Suppose, for example, we have two children, of different genotypes (i.e. arrays of genes). Under the environmental circumstances represented, let us say, by a contemporary well-off Swedish home, A grows to be 5 cm taller than B. Transplant the same two genotypes to a peasant environment in southern India, growing fitfully in the face of recurrent famines, chronic infections, physical labour and heat. Both children will probably end up smaller. But what non-linear interaction tells us is that B is unlikely to be still 5 cm smaller than A. He may be the same height or even taller in this different environment, for his genes may be more suitable for the regulation of growth in marginal circumstances.

Some interactions are more critical and dramatic. When a greylag goose hatches from its egg it has an inborn reaction to follow the first large moving object it sees. Normally this is the mother. But, as everyone knows, it will follow Dr Konrad Lorenz just as well if he should creep by first, and it will for ever after treat him as mother. The distinction between environment and heredity is a man-made one and evolution makes or ignores it according to the laws of Darwin rather than Aristotle. Animals are born into expected environments; some occupy niches of exquisite precision and negligible width; others, like man, are relatively polyvalent and can interact, with varying success, with a wide variety of environments. The classic work of Hubel and Wiesel and their successors (*see* p.116) provides a mammalian example of an essential interaction of genotype and environment. If one of an adult cat's eyes is covered for a few weeks

119

and then the cover is removed the cat is a bit clumsy for some minutes, then quite normal. But if one of a kitten's eyes is similarly covered for as little as a week from the time it normally begins to use the eyes, then the kitten never develops proper binocular vision; the cells of the cerebral cortex concerned with vision by the covered eye fail to develop their full characteristics. The genes have brought the cells to the point in time where light to both eyes is expected. Then the interaction must occur, or disaster follows. The experience of binocular seeing is necessary during the critical or sensitive period.

The facts of interaction lead to a principle from which the reader may deduce, if he wishes, a whole social philosophy. Everyone has a different genotype. Therefore, for optimal development (optimal, that is, for the individual and his environment considered individualistically), everyone should have a different environment. This is of course impossible; but one may seek either to optimize the interaction or to minimize it. In relation to cattle and other domestic animals we minimize the variation in both genotype and environment, aiming at a uniform product of maximal usefulness to ourselves. Such a strategy is foreign to successful evolution, and is fraught with the greatest danger. Should ecological circumstances change there is no reserve of genotypes suited to the new circumstances, no way forward and no way back. The true patrimony of a species is its genotypic variation.

So much for general principles. As we have seen, it is DNA, not stature, nor blue eyes nor even blood group A that is inherited. The sequence in the cell goes DNA$\rightarrow$ RNA $\rightarrow$ amino-acid assembly $\rightarrow$ protein. The protein escapes from the cell to influence another cell in building a tissue or organ; the organ interacts with others in the embryo and foetus; the foetus interacts with the uterine environment; the child interacts with the complex and changing environment of the adult-created and self-created world. If we talk simply about cellular proteins, the steps from genotype to character are small and well protected. In everyday language one may certainly say that the blood group A protein is inherited. But beyond this the going is not so simple. It is a long way from the possession of certain genes to the development of a height of 2 metres, say. For such a character the only valid statement is thus: 'x per cent of the variation in the height of young adults brought up in the circumstances of comfortably-off urban homes in the temperate climate of central Holland in the 1980s is due to genotypic variation'. For similar young adults brought up in Saharan African villages x would be less, for the environment would intervene more decisively.

Figure 37 shows a pair of identical, monozygotic twins, studied through the courtesy of Mr James Shields of the Department of Psychiatric Genetics of the University of London Institute of Psychiatry. The twins were separated at birth. One was brought up in a normal manner in a sympathetic home, while the other was watched over by a neurotic and cruel relative who shut him for long periods in the dark and required permission to be sought (and frequently refused) even for a drink of water. The two were brought together because of a search by Mr Shields for separated twins. Naturally, the deprived twin was more nervous than the other, but here we are more concerned with physical size, shape and body composition. The difference in size was considerable, amounting to 8·3

Figure 37   Monozygotic twins brought up apart under widely different circumstances from birth (*see* text)
(Twins seen by courtesy of Mr James Shields, 1962)

cm (or about 3ins) in height. In skeletal shape, however, the difference was negligible; subtract the fat of the healthy twin and enlarge the photograph of the deprived, and the monozygosity is clear to see. It must be said that this is an extreme and highly unusual example of a size difference in monozygotic twins, but it shows dramatically the effect of continuous childhood undernourishment and neglect better, perhaps, than would a large series of individuals contrasted with unrelated controls. (The only caveat is that we do not know the birthweights; part of the persisting size difference may derive from differences in intra-uterine growth.)

In the remainder of the chapter we shall necessarily leave this holistic approach and discuss in order:

1   Genetics of size, shape and tempo of growth
2   Effects of nutrition on growth and tempo of growth
3   Differences between races
4   Climatic and seasonal effects on growth
5   Effects of diseases
6   Psychosocial stress
7   Effects of urbanization
8   Effects of socioeconomic status and numbers in the family
9   Secular trend
10 Epidemiological auxology: height as a proxy for health

1. Genetics of Size, Shape and Tempo of Growth
Factors affecting the rate or tempo of growth must be considered separately from factors affecting the size, shape and body composition of the child. The genetical control of tempo seems to be independent of the genetical control of final adult size, and, to a large extent, of final shape. Equally, environmentally-produced changes in tempo do not necessarily affect final size or shape. Indeed, size and shape themselves seem to be separately controlled, both by genetical and by environmental factors. The genetical control of shape is more rigorous than that of size, presumably because shape represents chiefly how the cells are distributed, while size represents more the sum of sizes of the various cells. As explained in Chapter 2, the number of cells is fixed early, in the relative security of the uterus; their size continues to alter during much of childhood, and in some instances, such as the fat cell, throughout life.

Monozygotic twins, who have the same genotype, usually resemble each other very closely indeed when brought up under similar circumstances. Table 1 shows the average differences in length or height from birth to 4 years between pairs of monozygotic and dizygotic same-sexed twins, from the magnificent Louisville data of Wilson. Dizygotic same-sexed twins resemble each other genetically no more closely than any other brothers or sisters, since they arise each from a different fertilized ovum. At birth the monozygotic pairs were actually less alike

TABLE 1   *Mean differences between lengths of monozygotic twin pairs (~ 140 pairs) and same-sexed dizygotic twin pairs (~ 90 pairs) from birth to 4 years, and within-pair correlation coefficients*

	Mean difference in length (cm)		Correlation coefficient	
	MZ *pairs*	DZ *pairs*	MZ *pairs*	DZ *pairs*
Birth	1·8	1·6	0·58	0·82
3 months	1·4	1·6	0·75	0·72
6 months	1·3	1·9	0·78	0·65
1 year	1·3	1·8	0·85	0·69
2 years	1·1	2·4	0·89	0·58
3 years	1·1	2·9	0·92	0·55
4 years	1·1	3·2	0·94	0·60

(From Wilson, 1976)

than the dizygotic, but this situation rapidly changed. The increasingly close similarity of the values in monozygotic twins reflects size, shape and tempo combined. The differences between pairs at birth are probably partly due to asymmetrical division of the original ovum, one twin getting just a little more cytoplasm than the other, and partly to their different positions in an overcrowded uterus. The twin who is smaller at birth, by however little, usually remains smaller throughout life. When there is a large difference in size at birth, indicating some real intra-uterine distress to the smaller twin, the differences later are usually still quite small, so strong is the recuperative force. In Wilson's large-scale study of twins 10 monozygotic pairs differed by more than 150 gm at birth. All the same their average difference at age six was only 1·8 cm in height, 2·2 kg in weight and 1·6 points in IQ. On occasion bigger differences do persist, however, as in a small series reported by Babson and Phillips in Portland, Oregon.

The degree to which height is controlled by genotype when environmental circumstances are adequate is reflected in the variation within families compared with the variation amongst a population. The range of variation in adult height, represented by ± 2 standard deviations around the mean, is about 25 cm for most male populations, 16 cm amongst brothers and 1·6 cm amongst monozygotic twins brought up together. Height is generally said to be controlled by many genes, each of small effect. Such small-effect genes are called 'polygenes' and are believed to be located in parts of the chromosomes distinct from those carrying 'major' genes. The notion that height was controlled by polygenes arose because of the continuous and Gaussian distribution of height in the population (*see* Chapter 1, p. 19). It is true that the summation of many small effects does indeed bring about exactly this type of continuous distribution; but it is also true that a distribution virtually indistinguishable from it may be brought out by as few as five or six major genes interacting. The question is thus an open one.

Little is known about the genetics of shape. Some body measurements show higher correlations between parents and grown-up children than others. In a study of 125 Belgian families, containing 282 grown-up children, Professor Charles Susanne, of the Free University of Brussels found the correlations partly displayed in Table 2. In height each parent contributes equally to each offspring; despite all popular belief to the contrary, there is no tendency for daughters to resemble their mothers and sons their fathers; the four correlation coefficients (0·52, 0·47, 0·52, 0·53) do not differ significantly. Other data fully bear out this generalization. Farther down the table, the correlations for bi-iliac diameter (hip width) closely resemble those for height. But the biacromial diameter (shoulder width) values are less similar. The father–son correlation is very low, and the father–daughter is lower than mother–daughter. Amongst the head measurements, head breadth resembles height in pattern though all the correlations are rather lower. But in head length the father's correlation seems especially low, and in nose breadth brothers resemble one another not at all.

Interpretation of these correlations is not entirely straightforward. First, all the correlations are lowered somewhat if there is assortative mating in the population, that is if tall men preferentially marry tall women and short men short women. At least in industrialized societies, this occurs to the extent of producing a correlation of about 0·25 between heights of spouses. The relationship may not always be a straight-line one; at the two ends of the distribution the association gets closer as short men struggle to find shorter women and tall women look for still taller men (McManus and Mascie-Taylor, 1984).

Their measurements strongly affected by environment have low correlations; relatives simply do not resemble one another very much. If

TABLE 2 *Correlations between measurements of parents and grown-up children: 125 Belgian families*

Measurement	Parent–child	Brother–brother	Sister–sister	Father–son	Mother–daughter	Father–daughter	Mother–son
Height	0·51	0·53	0·57	0·54	0·47	0·52	0·53
Arm length	0·49	0·37	0·51	0·47	0·57	0·53	0·39
Sitting height	0·37	0·21	0·35	0·41	0·39	0·29	0·38
Biacromial diameter	0·33	0·42	0·46	0·09	0·50	0·33	0·41
Bi-iliac diameter	0·49	0·49	0·45	0·51	0·53	0·43	0·49
Head breadth	0·35	0·37	0·32	0·42	0·33	0·41	0·22
Head length	0·28	0·36	0·44	0·18	0·40	0·17	0·37
Nose length	0·31	0·00	0·44	0·32	0·26	0·34	0·35
Interpupillary breadth	0·38	0·34	0·40	0·32	0·40	0·42	0·38
Ear height	0·31	0·26	0·43	0·24	0·33	0·32	0·39

(From Susanne, 1975)

a measurement is chiefly controlled by a few genes of which one or more
shows dominance (i.e., a single dose gives the same effect as a double
dose), the correlations decrease. The presence of sex-linked genes (i.e.
genes on the X chromosome) causes the sister–sister correlation to exceed
the brother–brother correlation; the father–daughter to exceed the
father–son; and the mother–son to exceed the mother–daughter. No
example of this pattern appears in the table. What is clear is that different
measurements show different degrees and patterns of familial resemblance.
These differences cannot be understood until the development of each
measurement is studied and the physiological factors controlling its
growth are clarified.

Not all genes are active at birth. Some are not switched on till later, and
the products of others can express themselves only in the physiological
surroundings provided by the later years of growth. Some genes only
produce their effect in one sex, usually because the gene product needs the
co-operation of either one or the other hormonal environment in order to
exert its action. These genes are called 'sex-limited'. The effect may not
be all-or-none; the type of central baldness that is quite common in men
depends on a gene which in single dose causes an effect in males only. In
a double dose, which rarely occurs, it produces a similar effect in women.

Sex-limited genes are quite distinct from genes that are sex-linked. The
sex-linked genes affect alternately males and females in successive
generations (so-called 'criss-cross inheritance'). Quite often in sex-
linkage the affected males are worse than the affected females. This is
because the females have two X chromosomes and therefore two of the
genes in question, and one of them will be normal. Males have only one
X chromosome and thus lack the single normal gene.

The inability of some genes to express themselves till after birth
probably accounts for the fact that the resemblance in size between
children and their parents is quite small during the first 12 to 18 months.
The correlation between parents' heights and children's heights (or
lengths) is only about 0·2 when the children are newborn but swiftly rises
till it reaches its adult value of about 0·5 when the children are aged
about 18 months. This means that during the first 18 months many babies
change their centile positions for height and weight. Children with genes
making for large size but born to small mothers move upwards through
the centiles, and children born large but with genes making for small size
move downwards. Professor David Smith and his collaborators (1976)
give practical examples of this movement in babies in a well-nourished
American middle-class population; the upward shift was mostly complete

within 6 months after birth, while the downward shift took up to 18 months to accomplish. From age 2·0 till adolescence the parent–child correlation can be used to make standards for children's heights which allow for the heights of their parents. These standards, described in detail in Chapter 11, provide a more accurate means of diagnosing short stature than the ordinary population standards.

The correlation coefficients between height measurements of a child at successive ages and his own measurement as a fully-grown adult describe the curve shown in Figure 38. The correlation of length at birth with adult height is only about 0·3 since size at birth reflects uterine conditions much more than foetal genotype. Thereafter the correlation rises steeply and by age 3 is of the order of 0·8. This means that the adult height can be predicted from height at age 3, with an error which may amount to as much as 8 cm either way. (It takes a very much higher correlation than 0·8 to guarantee

Figure 38   Correlations between adult height and heights of the same individuals as children. Sexes-combined lines (0–5) from 124 individuals of Aberdeen study with + points from Berkeley Growth Study, California. Boys' and girls' lines (3–17) from 66 boys and 70 girls of California Guidance Study. All data pure longitudinal
(From Tanner, 1962, p. 88, where sources of data are detailed)

good prediction in the individual case.) At puberty, the correlations diminish because of some children maturing early and some late, but if bone age is taken into account the prediction is restored. Tables of prediction of adult height from childhood height and bone age are discussed in Chapter 11.

The correlation coefficients between height of the parents and height of the child at successive ages describe curves very similar to those shown in Figure 38.

Tempo of Growth   The genetical control of tempo of growth (as contrasted to shape and size) is well shown by the inheritance of age at menarche. Monozygotic twin sisters growing up together under West European conditions reach menarche some 2 to 3 months apart whereas dizygotic twins differ on average by around 10 months. The sister–sister and mother–daughter correlations are about 0.4. Thus a large proportion of the variability in age at menarche under these conditions is due to genetical influence. It is thought that mother and father exert an equal influence on tempo of growth, so that a late-maturing girl is as likely to have a late-maturing father as she is to have a late-maturing mother. In a study of a large sample of Swedish twins followed through puberty, Dr Siv Fischbein found the average difference between pairs in age at peak height velocity was five months for monozygotics and 10 months for dizygotics. The ages at which pairs of monozygotics reached the various pubertal stages were also highly correlated.

This control of tempo operates throughout the whole process of growth, for skeletal maturity at all ages shows the same type of family correlations as menarche. The age of eruption of the teeth is similarly controlled.

Under reasonable environmental conditions the genetical control extends down to many of the details of the growth curves. This is demonstrated by the records of three sisters shown in Figure 39. In the left-hand figure the heights are plotted against chronological age and in the right-hand figure against developmental age, as given by years before and years after peak height velocity at adolescence. Two of the sisters have curves which are practically superimposable except that they are on a different time-base, one being almost a year in advance of the other. These two therefore differ radically in one parameter of their growth curve, but little in their other parameters. The third sister differs little from the other two in velocity (when plotted against developmental age the curves are 'parallel'), but markedly in absolute height.

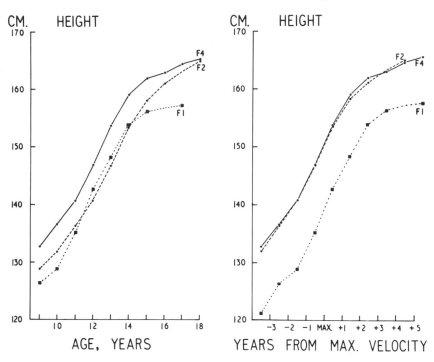

Figure 39   Growth in height of three sisters: (*left*) height plotted against chronological age; (*right*) height plotted against years before or after year of peak height velocity. Note the coincidence of curves of F 2 and F 4 when equated for developmental age (From Tanner, 1962, drawn from data of Ford, 1958)

2. *Effects of Nutrition on Growth and Tempo of Growth*

Malnutrition delays growth, as has been repeatedly shown by the effects of famine associated with war. Figure 40 shows the heights of children in Stuttgart, Germany, plotted at each year of age from 1911 to 1953. There is a uniform increase at all ages from 1920 to 1940 (*see* discussion of secular trend, p.156 *below*) but in both world wars the height dropped as the food intakes of the children became restricted.

Children subjected to an episode of acute starvation recover more or less completely by virtue of their regulative powers, provided the adverse conditions are not too severe and do not last too long. Chronic malnutrition is another matter. Most members of some populations, and some

Figure 40   Effect of malnutrition on growth in height. Heights of Stuttgart school children (7–8 to 14–15 Volkschule; 15–16 up, Oberschule) from 1911 to 1953: (*a*) boys; (*b*) girls. Lines connect points for children of same age and express secular trend and effect of war conditions.
(From Tanner, 1962, drawn from data of Howe and Schiller, 1952, and personal communication)

members of all populations, grow to be smaller adults than they should because of chronic undernourishment during most or all of their childhood.

We should distinguish nutritional effects on tempo of growth, on final size, on shape and on tissue compositon. Tempo seems usually to be the first thing affected; the undernourished child slows down and waits for better times. All young animals have the capacity to do this; in a world where nutrition is never assured, any species unable to regulate its growth in this way would long since have been eliminated. Man did not evolve in the supermarket society of today, but in small tribal communities, most of the period nomadic, following a usually precarious food supply. (Hence we can cope with periodic malnutrition better, perhaps, than with overfeeding.)

Adult size is affected by a less severe level of undernutrition than adult shape. Indeed, undernutrition in man does not alter shape very much, though there is evidence that the secular trend towards a larger size (*see below*) involves a change to relatively longer legs in relation to trunk. There is evidence also that at least in some societies, the difference in height between rich and poor is more due to leg length than to length of the trunk (Billewicz *et al.*, 1983). The traditional representations of Don Quixote and Sancho Panza may be but caricatures of a social reality.

In many populations the period when the child is most at risk from malnutrition, often combined with infection, is 6 months to 3 years. Though in some developing countries weights at birth are already low, in many (especially African countries) it is only after the first 6 months that weight-gains diminish. This often coincides with weaning and the substitution of high-starch, low-protein foods. It is also the age at which the mother's lactating ability declines so that satisfactory growth cannot be achieved by breast milk alone.

Professor John Waterlow of the London School of Hygiene and Tropical Medicine, and his colleagues there and in Jamaica, have made estimates of the energy requirements of infancy. Energy is measured in joules; the energy used simply for bodily maintenance in a child aged 1·0 years averages 330 kilojoules per kilogramme of body weight per day. The energy required for normal growth at this age averages about 20 kJ/kg/day. Normal infants use up some 80 kJ/kg/day in physical activity. Thus the maintenance requirement is a surprisingly high percentage of the whole. When calorie intake falls below 330 kJ/kg/day, as it commonly does in children in developing countries, then growth ceases. Even before this the energy margin for physical activity is eroded, and in the infant and child the restriction on exploration, play and social interaction that follows may be a more potent cause of delay in intellectual and emotional development than any nutritional effect on the nervous system.

Protein intake seems less important than was once thought. A child aged 1·0 years needs about 1·3 g milk protein per kg body weight; of this seven-eighths is used for maintenance and one-eighth for growth. If intake falls below this, growth ceases in 1 to 2 days. In the first 6 months, when the growth rate is higher, about three-eighths is used for growth. The protein–energy ratio in breast milk is about 7·5% in the first few weeks, falling to 5% at 2 to 3 months. This seems to be about the ratio required for healthy growth in childhood, and it is supplied by most foods, including cereals.

Figure 41 shows Dr Margaret Janes's work on the growth in height of two groups of boys in Ibadan, Nigeria. One group was drawn from the professional, highly-educated classes, living in considerable affluence;

Figure 41   Average growth in height of two groups of boys in Ibadan, Nigeria, plotted on British Standards: (*a*) well-off group; (*b*) indigent group
(Data from Janes, 1975, as quoted in Eveleth and Tanner, 1976)

the second from indigent slum-dwellers in the market area of the town. The curves have been drawn on British standard charts. The well-off grow very similarly to British boys, while the 50th centile for the poor is no higher than the 10th centile of the British. There is no evidence that the two groups differ much in genotype, though they differ in other things besides nutrition. However, the major cause of the great difference in growth curves is certainly nutritional level.

Nutrition and infection   The most important of the adjuvent causes of poor growth is undoubtedly infection, and particularly diarrhoeal disease. Dr Raymond Martorell and his colleagues showed that in Guatamalan villages, where 50% of the children aged three or less were sick on any given day, those who had been relatively free of diarrhoea averaged 3·5 cm more in height, at age seven, than those with a greater number of diarrhoeal episodes. Similar results were obtained in a detailed study in the Gambia (Roland *et al.*, 1977). Children with diarrhoea take in less food, a deficit aggravated in some cultures by deliberate restriction of diet when a child is sick. Other infections may have the same effect: Duggan and his colleagues carried out 24-hour energy balance studies in 20 Kenyan children during acute attacks of measles, with control periods after recovery. The energy intake fell by 75% during the attack, and the children went into negative nitrogen balance, to the tune of 169 kJ/kg per day on average. If such attacks are followed by a period of increased nutrition, to provide for the necessary catch-up, well and good. But if not, then the effects of undernutrition will follow. In such a situation simply giving extra food throughout childhood is far from a complete answer. At high levels of infection it will be ineffective, and even at moderate levels the extra amount has to be considerable at the required time. The infections must be eradicated too. In some third-world situations the simple provision of well-trained medical personnel dedicated to primary care, advice and prevention can of itself lead to dramatic results. The arrival of a Medical Research Council team in a group of Gambian villages caused, in eight years, a drop of the infant mortality rate from 149 to 24 and of the early childhood rate from 109 to 23 (Lamb *et al.*, 1984). Undoubtedly the growth rates in such a situation would be rising sharply; and after these relatively low mortality rates had been achieved, growth rate monitoring would move to the centre of the picture, as further effects on mortality become harder and harder to demonstrate.

In a more recent study of a small coastal town in the Gambia, Dr Roland and his colleagues (1988) examined each month one-third of all the babies born during the year. All had free access to a paediatrician as required. As is quite usual in developing countries, the infants' weights were above those of the American standards until 6 months, but both weight and height velocity fell below the standard at 3 to 4 months, so that by one year the children were quite small. The occurrence of infections was carefully logged for each infant. The amounts of both diarrhoeal disease and upper respiratory infection correlated negatively with weight gain; diarrhoea accounted for half the weight loss *vis-à-vis* the standards, respiratory disease a further quarter. The impact of diarrhoea was mostly in the seven to twelve month period, and was four times as great after weaning as before. Effects on length in this cohort were negligible.

Effects of food supplements   A number of studies of the effects of giving food supplements have been done, both in developed and undeveloped countries. Amongst the former, a particularly careful trial was carried out by Dr Elwood and his colleagues in two small towns in South Wales. A consecutive series of 1251 pregnant women were allocated at random to a group supplemented with extra milk, and a control. The milk was drunk by the mothers in the last four weeks of pregnancy and by the child (the modest amount of $1/4$ pint per day) from birth to age five. Other children in the family were also supplemented if under five. At age five no difference was demonstrated in height, weight, head circumference or skinfolds. Even in a socially 'vulnerable' subgroup with many sibs and low food expenditure no effect was seen. The sensitivities of the trial were such that a difference of 0·8 cm would have been detected at a 5% level of significance. Differences in mean heights between upper and lower social classes in the UK considerably exceed this amount. In the light of such feeding trials and of studies of dietary intake which fail to reveal any considerable social class difference (Wenlock *et al.*, 1986) it is hard to ascribe the social class differences in height existing in most Western industrialized nations simply to differences in nutrition (*see discussion below*).

In the underdeveloped countries, too, the same generalization holds. Supplements of 200 cals/day, 22 g protein, given to a group of mothers and poor children in Bogota, Columbia, during the last three months of pregnancy and the first three post-natal years did result in increased growth compared to controls; but not to the extent that the heights and weights reached those of the local rich children. The incidence of infection remained unchanged. Mora and his colleagues write 'Simple provision of food supplements without effective prevention of infection and particularly diarrhoeal disease is likely to have limited effects on the prevention of physical growth retardation amongst children'.

Adult outcome of childhood undernutrition   The question whether undernutrition during foetal life or during the first 1 or 2 years after birth necessarily leads to a deficit in size or in mental function in adults has been frequently and sometimes heatedly discussed. It seems that children with severe protein–calorie malnutrition in early infancy due to malformations or malfunction of the gut (*see* Valman, 1974) make a complete recovery in height after surgical correction, when brought up in reasonably well-off homes in a developed country. In the majority of studies, they have also caught up completely in intellectual ability, when assessed at ages 7 to 10 years. In one study, however, it seemed that starvation in the first 3 months after birth might have been responsible for a lowered score in auditory memory tests, though not in tests of general intelligence.

Evidently much depends on the circumstances after the severe episode of malnutrition is over. Children under 5 years old admitted to hospital in tropical countries with severe malnutrition (kwashiorkor or marasmus) have been followed up after leaving hospital. In most such children complete equality of height and weight with sib controls has been attained before puberty. In trying to isolate the effects of malnutrition from general social effects, sibs are more satisfactory controls than the general population, since they suffer some at least of the same environmental circumstances. Naturally, the children admitted to hospital come from poorly-off families where even the sibs are growing relatively badly.

Dr Steven Richardson of the Albert Einstein College of Medicine, New York, sums this work up very precisely in two generalizations: 'Where the histories of both the severely malnourished child and the comparison child suggest a similar level of nutrition over their life histories except for the presence or absence of an acute episode requiring treatment in hospital, and where both sets of children have experienced similar social, physical and biological environments, the children with an acute episode of malnutrition will not be smaller in somatic growth [at school age] than their comparisons' and 'Where the children with an early acute episode of malnutrition are more disadvantaged than their comparisons in the overall histories of malnutrition, social, physical and biological environments, then the malnourished children at school age will be smaller in somatic growth than their comparisons'. He adds, however, that his own work suggests that the second generalization needs qualification: 'Even when the children with acute malnutrition come from environments and backgrounds that are somewhat disadvantaged, complete catch-up may still occur' (Richardson, 1975).

A very well-controlled study of the long-term effects of undernutrition during foetal life was made by Z. Stein and M. Susser and their colleagues at Columbia University, New York. Between mid-October 1944 and May

1945 the central part of Holland, including the towns of Rotterdam, Amsterdam and Leiden, was subjected to a severe war-time famine. The official daily ration was 1,500 calories and the actual amount eaten only slightly more than this. Examination of the birth records in these three towns by monthly cohorts of birth showed that birthweight diminished by 9% in the cohorts who were exposed to the undernutrition in the third trimester of pregnancy (i.e. from 6 to 9 months). Babies exposed only in the first or second trimesters showed no weight reduction. Birth length was reduced by 2·5% and head circumference by 2·7% in the third trimester cohort. Maternal weight was 10% less than the control values taken after May 1945.

At age 19 the males of this group of children entered military service. At entry their height was measured and several tests of mental ability were administered. In addition the incidence of mild and severe mental retardation was ascertained for the relevant cohorts. Stein and Susser found that young men who had been undernourished in foetal life were no different in height, weight or mental ability from those who had not been undernourished. Nor was the incidence of mild or severe mental retardation any greater. They searched for differential effects of the famine exposure according to social class, family size and birth order, but found none. Their final conclusion is an impeccable summary of the present position: 'We believe we must accept that poor *prenatal* nutrition cannot be considered a factor in the social distribution of mental competence among surviving adults in industrial societies. This is not to exclude it as a possible factor in combination with poor *postnatal* nutrition, especially in pre-industrial societies' (*Famine and Human Development*, p. 236, their italics).

Experimental models   A considerable amount of experimental work on animals has been done, designed to throw light on the effects of long-term undernutrition. As a result many and sometimes strident claims to certain knowledge have been made, especially in relation to the highly important and emotive subject of brain growth. Such claims should be treated with the utmost caution; our knowledge in this area at the present time is not at all great.

An early difficulty was the *naïveté* of physiologists about species differences. Much of the early experimental work was done on the rat, a species whose growth is about as different from man's as it is possible to find amongst the whole range of mammals. Rats are born at a far earlier stage in development than primates, at an age roughly corresponding to 16 postmenstrual weeks in man. Thus, undernutrition that is immediately postnatal in the rat corresponds, if to anything in man, to undernutrition of the foetus in mid-pregnancy. This does not result from maternal undernutrition (unless at catastrophic levels) but only from pathology of the placenta, a wholly different situation. Besides this, the rat bears litters not singleton foetuses, its whole endocrine system develops in a very different way from that of man, and there are enormous differences in behaviour and brain function. The rhesus monkey and the New World cebus are much more suitable models. Dr Donald Cheek of Melbourne, Dr Don Hill of Toronto and Dr George Kerr and his associates of the Harvard School of Public Health have in particular studied the former, and Dr Hegsted and his co-workers also at the Harvard School of Public Health, have investigated the latter (Fleagle *et al.*, 1975). Primates all have the same characteristic growth-curve, although monkeys are born at a more developmentally advanced stage than apes, and apes at a more developmentally advanced stage than man. Endocrine development in monkeys is not dissimilar from man's and even the brain shares many of man's preoccupations.

Cheek and Hill studied malnutrition during the foetal period. The rhesus monkey (*Macaca mulatta*) has a pregnancy lasting 165 days from fertilization. Hill tied off some of the placental vessels at 100 days to cause relative placental failure and intra-uterine malnourishment in the foetuses. At birth the animals were severely underweight, averaging about 2·5 standard deviations below the mean of controls.

Cell numbers and cell size in the cerebrum were entirely unaffected; in the cerebellum a small reduction in the number of cells occurred. In this species the maximum velocity of cerebrum growth takes place from 80 to 100 days after fertilization, i.e. before the experimental intervention. Thus the negative result is quite expected. In relating this to man, however, one should remember that the rhesus at birth has a brain weight 60% of the adult value whereas man's brain develops more slowly and at birth is only about 25% of the adult value. From the cerebrum-time-table point of view, the placental insufficiency studied by Hill corresponds to placental insufficiency near to birth in man (judging by the timing of maximum growth rates of brain weight curves). The cerebellum develops later than the cerebrum in both species (*see* Chapter 8), which probably explains the slightly greater effect of the malnutrition upon it.

The work on cebus has been in the postnatal period. The monkeys were separated from their mothers at birth and reared in a primate nursery with other monkeys. For the first 8 weeks they were fed a commercial human baby food and thereafter a synthetic liquid formula of which the control monkeys had as much as they wished. Undernutrition began at this time in two experimental groups, one simply fed 67% of the controls' diet and the other fed a very low-protein diet (2 8% of calories as protein) but in as large a quantity as desired. The three groups were followed for 20 weeks, then the two experimental groups were rehabilitated, eating as much of the control diet as they wanted. By the end of the 20-week experiment period the controls had doubled their weight, the calorie-deficient had increased their weight by 20% and the protein-deficient had scarcely gained weight at all.

Extensive physical measurements were made on these monkeys so that changes of shape as well as size could be followed. The growth of all parts of the skeleton slowed down, and did so to a greater extent in the protein-deficient than in the calorie-restricted group. Some growth occurred in nearly all skeletal measurements, however, even in the worst-off group. The shape changes could be important, theoretically for throwing light on the mechanism of growth, and practically as an aid in diagnosing undernutrition in human infants. J. G. Fleagle, D. M. Hegsted and their collegues found that some skeletal relationships stayed quite constant under these conditions; e.g. the relative lengths of upper and lower arm. But feet and hands suffered more than legs and arms, and sitting height more than lengths of arms and legs. It seems that those dimensions which were farthest along their road to maturity were held up most (hands are ahead of forearms, feet ahead of calves, sitting height ahead of limb lengths, in cebus as in man; see Chapter 10, p.175). 'In response to malnutrition', write the authors, 'the available resources are expended disproportionately in favour of parts with the greatest distance to cover in the future . . . in long-term malnutrition such a distribution would tend to lead to a closer approximation of "normal" adult skeletal proportions than would a more uniform reduction of growth rates' (Fleagle et al., 1975). The second point is an important one: the arrangement is adaptive, in the biological sense of preserving the competitiveness of the species in the struggle for reproductive success. If we assume that the normal shape of cebus adapts it optimally to its ecological niche, then the response to chronic undernutrition, a circumstance clearly encountered very frequently, should be such that the advantages of normal morphology are preserved as closely as possible. What is sacrificed is size, what is preserved is shape. The same is seen in human malnutrition.

The results of rehabilitation have also been reported. Catch-up growth in weight was rapid and complete in both groups, and by 22 weeks of rehabilitation (1 year of age) limb catch-up was complete in the calorie-deficient group and very nearly so in the protein-deficient group. Clearly size and shape in the end returned to normal. These experiments had another and highly instructive side. In real life undernutrition of human infants almost invariably occurs in combination with social deprivation in conditions of poverty. Hence, when such infants are treated in hospital and returned to their families and re-tested months or years later, it is extremely difficult to know whether any deficit in intellectual or emotional functioning is due to the malnutrition itself or to the effect of the associated social circumstances. Even comparison with sibs who did not enter the hospital raises the obvious problem of whether one child is favoured over another.

The Harvard investigators tackled this problem by dividing the cebus infants on restricted diets each into two groups. One group was reared in partial isolation, in a solitary cage within sight and sound of other monkeys; and the other group was housed similarly but had 3 hours of play every weekday with 1 to 3 other monkeys in a large play-pen, plus daily human handling. The monkey analogues of intelligence and emotional poise in humans are perhaps not easy to define; exploratory behaviour was the item chiefly studied, plus motor behaviour, general activity and the occurrence of items such as rocking, clutching of the body and frowning, which signify emotional disturbance.

By the end of the 20-week period of food restriction it was clear that well fed and socially deprived animals scored about the same as ill fed and socially enriched ones, both groups scoring lower in all items than the well fed and socially enriched. The authors, M. F. Elias and K. W. Samonds, concluded that their findings were consistent with many others in showing that the behavioural consequences of calorie malnutrition were in many respects similar to those of emotional deprivation. Simultaneous deprivation in both factors was more severe than deprivation in either alone, but the two factors were additive rather than interactive in effect. The retardation of behavioural development was less than that of growth in size, though not much different from the retardation in skeletal maturity. After six months rehabilitation, no behavioural deficit could be found in either of the two nutritional groups; recovery was complete.

From a practical point of view the outcome of the monkey experiments and of those human experiences which are carefully controlled seem to agree in suggesting that events subsequent to a period of malnutrition occurring either in late foetal or early post-natal life exert an overridingly important effect on its outcome. The restorative powers are very great provided social circumstances allow them full play.

Bigger not better   One last important point needs to be made here. It should not be uncritically assumed that bigger means better. Just because in temperate industrialized countries the better-off members of the community are larger than the worse-off and tall women have more successful reproductive histories than short women, it does not follow that the same holds true under other ecological conditions. Small body size may be more adaptive under some conditions. In an agricultural peasant economy a small man is more efficient than a tall one, requiring to do less work to feed himself. Level of nutrition has to be seen as part of the whole ecology, of a whole philosophy even. Overnutrition is no less lethal than undernutrition, and a great deal more prevalent in many parts of the world. As Stini (1980) writes 'The infant subjected to a year of *ad libitum* [bottle] feeding of high-protein milk or formula is the possessor of a substantial bulk of both lean body mass and adipose tissue, neither of which can be functionally justified. This is maximization, not optimization'. In fact, in the parts of the Western world dietary intakes in infancy diminished in the 1970s compared to the 1950s and 1960s, with a reduction in infant skinfolds (Whitehead and Paul, 1984).

3 Differences Between Races
Populations differ in their average adult size and their tempo of growth as well as in their shape. These differences, as we have seen, are due to a complicated interaction of genetical and environmental factors. In this section we concentrate on the effect of differences in gene pools represented by the rather arbitrary classification of race. We rely on geographical and historical origins for our criteria for these major population groups, and refer to them simply as European, African and Asiatic (by which last is meant Chinese, Japanese and Indo-Malayan). For the purposes in hand we want to compare representative groups of each race growing up, so far as possible, under similar and preferably optimal environmental conditions. Extensive relevant data may be found in the book *World-wide Variation in Human Growth* by Dr Phyllis Eveleth and the present writer (second edition 1990).

Figure 42 shows the height curves of London children, well-off Chinese children in Hong Kong, and Afro-Americans from Washington, DC, who, though from a relatively low-income group in the United States,

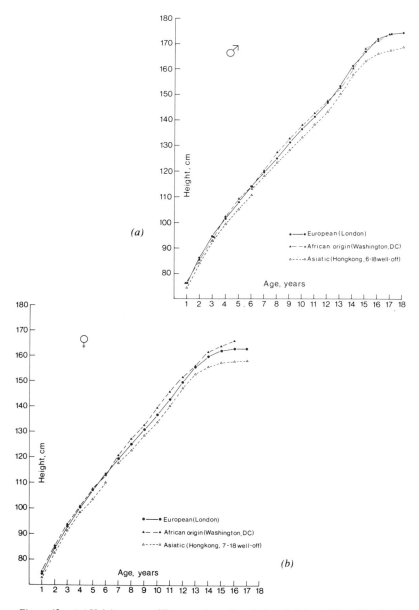

Figure 42 (*a*) Height means of European boys (London), Asiatic boys (Hong Kong) and USA boys of African origin (Washington DC); (*b*) Height means of European girls (London), Asiatic girls (Hong Kong) and USA girls of African origin (Washington DC) (From Eveleth and Tanner, 1976)

lived in a more favourable environment than any African group studied in large numbers in Africa (the Nigerian well-off group approach this condition, however; *see* p. 132). The European and Afro-American groups have almost identical curves for boys; the African-descended girls, however, are a little larger than the European girls. This is partly due to their having a swifter tempo, with menarche about 0.3 years earlier.

The Asiatic boys and girls are distinctly less tall, despite coming from high socio-economic groups receiving better-than-average care. It is not growth delay that makes them smaller; indeed their tempo of growth is significantly faster than that of Londoners, with menarche at 12.5 compared with 13.0 years. Hence the adult height difference is greater than the childhood difference, not less. This difference is due mainly to differences in gene pool. A frequently-quoted finding, due to Professor William Greulich of Stanford University, is that Japanese children growing up in San Fransisco are taller than Japanese growing up in Japan. This was indeed the case in the 1950s, but by 1980 the marked trend towards increasing height in Japan had eliminated the difference. Japanese in Japan, Hawaii and California no longer differ significantly in height; but both differ from Hawaians and Californians of European or African descent, their mean heights being at about the 15th centile of the British standards.

Bodily proportions   The largest differences between races, when all are growing up in good environments, are those of shape. Figure 43 shows lines representing sitting-height means at successive ages plotted against leg-length means at the same ages, in boys. The London boys (solid line) show a straight line up to adolescence, at which time sitting height spurts more than leg length, causing the line to bend upwards. The Chinese when small have a similar body proportion, but by adolescence their sitting height has become somewhat greater for a given leg length. (This however may not be true of the modern Japanese, *see* Secular Trend, *below*.) The Africans have a much lower sitting height for leg length (or, equivalently, longer legs for similar sitting height) and the Australian Aboriginals have still longer legs than the Africans. These are characteristic differences, shown by females as well as males (*see* Hamill and colleagues, 1973, and Greulich's paper). Figure 44 illustrates the European–African differences in relation to two typical Olympic sprinters.

An equally characteristic difference between European and African is in the relation of shoulder width to hip width. In both sexes the African has slimmer hips for a given shoulder width. The proportions of the Asiatic in

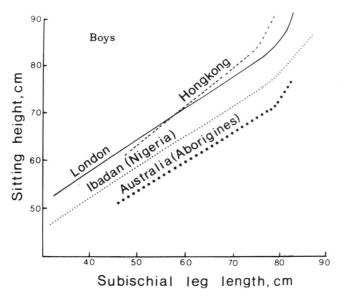

Figure 43 Regression of sitting height means on leg length means at successive ages of European, Asiatic, African and Australian populations
(From Eveleth and Tanner, 1976)

Figure 44 Comparison of European and African physiques with photographs enlarged so that both have the same sitting height: two Olympic 400 m runners
(From Tanner, 1964)

this respect are generally similar to those of Europeans, though there are recent data showing Americans of Mexican origin have broader hips and limb widths for given height than Americans of European origin (Martorell *et al.*, 1987).

There are differences in body composition also, Africans having more muscle and heavier bones per unit weight, at least in males, together with less fat in the limbs in proportion to fat on the trunk (Johnston *et al.*, 1974; Eveleth and Tanner, 1976). These differences confer distinct advantages and disadvantages in certain sports, with results that can be seen in Olympic records (Tanner, 1964). Africans have an advantage in many track events, especially the high hurdles; the Asiatics in gymnastics and weight-lifting.

Tempo of growth   The African newborn is ahead of the European in skeletal maturity and motor development. He maintains this advance only for some 2 or 3 years in most areas in Africa, but this is because nutritional disadvantage supervenes. In America and Europe the African stays in advance, in bone age and also in dental maturity. In a nationwide survey of the United States, in which a true proportionate sampling of the whole population was attempted, the mean ages of menarche were 12·5 years for African-descended and 12·8 years for European-descended. Well-off Asiatic groups have as fast a tempo as Africans, in later childhood if not in earlier years. Mean age of menarche in Hong Kong girls from affluent families was found by Low and his associates to be 12·4 years. In Japan in the 1980s age at menarche averaged 12·5 (Nagai *et al.*, 1980).

4.   Climatic and Seasonal Effects on Growth

It might be argued that the differences in growth that occur between Europeans, Africans and Asiatics might be due to the direct effect of climate on the growing child, were it not that we are able to compare descendants of Africans and Europeans living in the same area. The differences in growth have indeed been brought about in response to the different ecological conditions in which each group evolved, but they arise from selection over many generations rather than from the immediate effect on individual children. The long limbs of the African enable him to lose more heat per unit volume than the European, and the thick-set body and short limbs of the northern Asiatic are similarly adaptive in arctic regions. There is, in fact, quite a close correlation between the linearity of peoples, as judged by their weight per unit height as adults, and the average annual temperature of the area where they live, or from which they migrated in historical times. Thus over thousands of years children and adults with genes leading to climate-adaptive characteristics have survived to breed more frequently than others.

One direct effect of climate on growth is certainly known. The high altitude of the Peruvian *altiplano* (4,000 metres) induces a larger chest circumference and bigger lungs in Quechua children growing there than occur in Quechua children on the sea coast. The same occurs in children of European origin living in comfortable social circumstances in La Paz. Their height, however, is reduced, by an amount estimated at 4 cm (Greksa 1988).

One alleged climatic effect has achieved a certain notoriety; nine-teenth-century medical textbooks stated that menarche was earlier in girls living in the tropics, and this information has been duly copied from one textbook to another till quite recently. In fact, modern statistics make it clear that climate has at most a minor effect on age at menarche; people living in tropical countries mostly have, in fact, a late menarche, but simply because their nutritional level is low. Well-off children in these areas have menarche at approximately the same time as children in temperate zones, and children of temperate-zone parentage growing up in the tropics have menarche at a normal time for the population from which they are drawn. It is true that girls in Mediterranean countries such as Spain, Italy and Greece have an earlier menarche (12·1–12·8) than girls in comparable circumstances in north-west Europe (13·0–13·4) but this seems to represent a genetic difference. Australian first-generation descendants of both groups growing up cheek by jowl in Sydney sub-urbs, keep their distinctive tempos. (Jones, *et al.*, 1973). A list of median ages of menarche in different populations is given in Table 3.

Season of the year, however, exerts a considerable influence on velocity of growth, at least in west European children. Children grow faster in height in spring and summer than in autumn and winter. In England, Professor W. A. Marshall found that in children aged 7 to 10 years maximum height velocities were reached in the 3-month periods ending between March and July. During the 3 months of fastest growth a child on average grew three times as much as during his 3 months of slowest growth. Thus the difference is considerable and sufficient to make it important to study growth over a whole year when comparing the effect of a hormone or a drug on growth rates. Even in northern California where the temperature change is less, Lee (1980) found a similar, though smaller, effect.

TABLE 3   *Mean ages of menarche (years) in various population groups. All data refer to period between 1965 and 1985: status quo data with means calculated by probits or logits. Sources of data are Marshall and Tanner (1986) and Eveleth and Tanner (1990)*

Europe		Near East and India	
Reykjavik	13.1	Bagdad (well-off)	13.6
Oslo	13.2	Istanbul (well-off)	12.4
Sweden	13.0	Jerusalem	13.3
Finland	13.2	Tunis (well-off)	13.4
Denmark	13.0	Madras (well-off)	12.9
Netherlands	13.3	Punjab (well-off)	12.5
North-east England	13.4	Cairo (well-off)	12.6
London	13.0		
Brussels	13.1	*Asiatics*	
Paris	13.0	Singapore (rich)	12.4
Zurich	13.4	Hong Kong (well-off)	12.5
Moscow	13.0	Japan (urban)	12.5
Warsaw	12.7	Mexico City (well-off)	12.3
Hungary	13.1	Yucatan (well-off)	12.6
Romania (urban)	13.5	Eskimo	13.8
Carrara, Italy	12.6		
Naples (rural)	12.5	*Africans*	
Athens	12.6	Somalia (better off)	13.1
Madrid	12.2	Nigeria, Ibadan	13.3
		(university-educated	
European-descended		parents)	
Montreal, French	13.1	Khartoum (better off)	13.4
USA, all areas	12.8		
Sydney	13.0	*African-descended*	
New Zealand	13.0	USA, all areas	12.5
Argentina (urban)	12.5	Cuba, all areas	13.0
Chile (urban)	12.6		
Brazil (urban rich)	12.2	*Pacific*	
Venezuela	12.0	New Zealand (Maori)	12.7
(urban rich)		New Guinea (Bundi)	18.0
		New Guinea (Megiar)	15.5
		Tahiti	12.7

Marshall also showed that totally blind children failed to synchronize their growth rate variations with the time of year, though the variations themselves occurred to the same degree as in sighted children. The cause of the variations is unknown; but whatever it is, light or some radiation falling on the eye 'entrains' them to season of the year (as it entrains the breeding-cycle of many animals).

In tropical countries seasonal variations tend more to be governed by the rainy and dry periods, and follow food supply and the frequency of infections (Rosetta 1988).

5. Effects of Diseases

In well-nourished children the effects on growth of minor diseases such as measles, antibiotic-treated middle ear infection, or even pneumonia, are minimal. Hermanussen and his colleagues at Kiel, using the Valk knemometer, have shown that growth does sometimes slow down or even stop during the illness but an immediate catch-up fully restores the situation.

In children with less adequate diets the catch-up may not restore the situation so completely. Often children with continual colds, ear disease, sore throats and skin infections are on average smaller than others, but they tend to come from economically depressed and socially disorganized homes, where proper meals are unknown and cleanliness too much trouble. The relation of infection and nutrition has been discussed above. In the National Study of Health and Growth surveillance (see below), Rona and Florey found that children aged 5–11 with a history of frequent respiratory symptoms such as cough and bronchitis (but excluding asthma) were smaller than others only when they were in the lowest socio-economic group and had many siblings. Children with asthma are smaller on average than others, though this seems to be primarily or entirely an effect on tempo rather than final size.

One environmental pathogen has been clearly shown to have an effect on growth. This is smoking by the mother in pregnancy. There are a very large number of studies on this; a recent one in South Wales by Elwood and his associates (1987) estimated the effect of smoking more than 25 cigarettes a day in mid pregnancy as a birth weight lowered by 300 g, birth length by 1 cm, head circumference by about 0.5 cm. Like the small-for-gestational age effect, the smoking effect persists: in the Elwood data the length difference between the children of non-smokers and heavy smokers remained 1 cm at age 5. Actual disorders of the growth process are discussed briefly in Chapter 12.

6. Psychosocial Stress

There is now clear evidence that in some children psychological stress causes relative failure to grow (see Chapter 12). Quite apart from alterations in appetite, it may do this by inhibiting the secretion of growth hormone. When the stress is removed, secretion of growth hormone resumes again, and in clinical cases a catch-up occurs which is indistinguishable from the catch-up following administration of growth hormone to a child otherwise permanently deficient in it.

Just how far information derived from such clinical cases may be extrapolated into the life of the ordinary child is hard to say. Fairly severe stress seems usually to be involved in these clinical cases and, of course, the majority of children given sufficient food continue to grow even in astonishingly stressful circumstances. In a famous experiment (originally planned to illuminate an entirely different issue) Dr Elsie Widdowson of Cambridge University furnished evidence that the presence of a sadistic schoolteacher caused a slowing down of growth in children in an orphanage, despite a concurrent increase in the amount of food eaten. Presumably activity increased, unless indeed stress put up the requirements of maintenance. Many years ago Drs G. E. Friend and E. R. Bransby found that in certain boarding-schools boys grew more slowly in term-time than in the holidays at home. Unpublished data of R. H. Whitehouse and the present writer confirmed this. On the other hand, a boarding-school may provide the friendly atmosphere required for catch-up to a child whose growth has been stunted by an adverse home, as illustrated in Chapter 12.

7. Effects of Urbanization

Children in urban areas are usually larger and have a more rapid tempo of growth than children in the villages of the surrounding countryside. This apparently results from the provision of a regular supply of goods, of health and sanitation services, and educational, recreational and welfare facilities. Areas of high population density which lack such characteristics, as for instance the shanty slums of South America and Africa, do not show the same effect on growth as the orderly towns of the industrialized countries. In Finland, Greece, Austria, Romania and Poland, boys in the towns are between 2 and 5 cm taller (depending on age and the country considered) than boys in the villages. Some statistics on age at menarche are shown in Figure 45.

In Poland a particularly interesting study was made by Drs S. Panek and E. Piasecki when a new town was founded, with the immigrants all coming from a near-by rural area. Children growing up in the town were compared with children whose parents stayed in the rural area. They

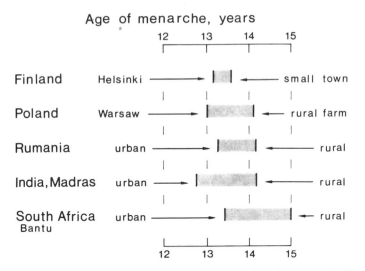

Figure 45   Mean ages of menarche in urban and rural areas throughout the world. Shaded areas show difference between median ages in each geographic area (From Eveleth and Tanner, 1976)

averaged 4 cm taller; in the girls menarche occurred earlier. It is possible that the migrants from country to town were a special selection of tall, fast-growing people. Such self-selection is characteristic of most migrant groups. It is a well-known difficulty in all studies of voluntary migrants that they never seem to be a random sample of the population they leave, but are usually taller and more intelligent than the stay-at-homes (even if they migrate only from one English county to another).

Various highly speculative reasons have sometimes been given for the swifter growth of urban children. They range from more continuous illumination with artificial light to more continuous bombardment with sexual stimulation. In fact, the nutritional explanation suffices: rural children expend more energy in physical activity than urban ones, and their total calorie intake is frequently less, despite townsmen's ideas to the contrary.

8. Effects of Socioeconomic Status and Numbers in the Family
Children from different socioeconomic levels differ in size and in tempo of growth in every society except two (p.154) so far examined. The upper groups, defined usually in terms of father's occupation, are larger and

have a more rapid tempo. In most European groups – both West and East – the chief difference is between those whose fathers are in occupations classified as 'manual' (or 'working') and 'non-manual' ('intellectual'). In developing countries similar or larger differences occur in relation to fathers' or mothers' educational status. The classification by occupation, as in the Registrar-General's classification in the UK, has been very useful for many epidemiological purposes, but is becoming increasingly irrelevant as an indication of the standard of living and the child-centredness of the home, which are probably the operative factors.

The difference in height between children of professional and managerial fathers and those of unskilled manual labourers in the UK is currently, in the case of boys, nearly 2 cm at age three rising to 4 cm at adolescence and falling again to 3 cm in young adulthood. In girls the differences are slightly lower; at three about 1·5 cm, adolescence 3 cm, and in adults 2 cm (*see* Rosenbaum *et al.*, 1985). Part of the difference, then, is due to a faster tempo of growth in the well-off. At least in the British context, much of the difference is established between ages one and three. In a study of 1,163 consecutively-born children in two small towns in South Wales, Elwood and his colleagues (1987) found no social class

Figure 46   Height at 7·0 years of British children (sexes combined) according to number of children in the family and socio-economic level (1–5)
(Data from National Child Development Study, 1965; given in Goldstein, 1971)

differences in length at three months but a manual/non-manual difference of 0.5 cm at 6 months rising to 1 cm by age 3.

In weight the differences are somewhat less, since the worse-off children have a higher weight for height partly, it seems, because of a higher intake of sugar and other carbohydrates relative to fat and protein and partly because, even though total energy intakes are not different (*see* Wenlock *et al.*, 1986), they take less exercise. (In the same vein the incidence of obesity is frequently greater in the lower social classes in developed countries.)

In the British National Child Development Survey (NCDS) – a nationwide sample of children consisting of all those born in the first week of March 1958 (*see* p. 163) – an overall difference of 3.3 cm in height was found between 7-year-old children of managerial and unskilled labouring fathers (Figure 46). Similarly in the US national sample of children measured by the National Center for Health Statistics in the 1960s children aged six to 11 were some 3 cm taller in the rich families than in the poor ones. In the 1970s, the HANES I and II surveys related height to a poverty index, compounded of income, family size, sex of head of household, rural or urban residence and some other factors. An index of 1.0 was regarded as the poverty line, that is a level just adequate to feed and clothe a family. Boys in families below this level averaged about 2.5 cm, and girls about 1.5 cm below those in families above it; this at ages five to 17 (Jones *et al.*, 1985; *see also* Garn *et al.*, 1984).

These differences are compounded by differences in height and tempo according to numbers of sibs in the family (Figure 46). As one might anticipate, first-born children are (during their childhood) somewhat taller than later-born children with the same number of sibs, since they have had a period of being an only child. Thus the effects of being a later-born child (for given age of mother) and of having younger sibs summate. The more mouths to feed, it seems, or simply children to look after, the more slowly the children grow. In addition, the more older sibs a young child has, the more likely he is to catch infections, and to catch infections younger than he otherwise would. Reves has emphasized, indeed, that the fall in sib number may be the greatest cause of the decrease in preschool child mortality in England in the period 1900–1950. Professor Harvey Goldstein of London University Institute of Education, in a now classic paper, showed that in the NCDS data about half of the social class differences in height persisted even when the effect of numbers of sibs was allowed for. The more recent data of the National Study of Health and Growth are illustrated in Figure 47. A marked non-parallelism is seen

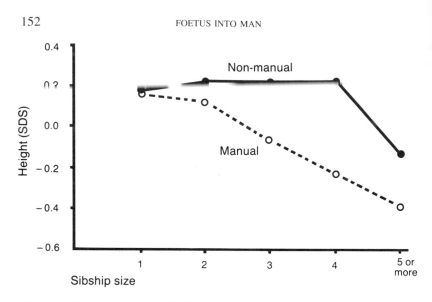

Figure 47   Height of 5–11-year-old children in England expressed in Standard Deviation Scores (SDS) according to occupational status of father and sibship size (From Rona *et al.*, 1978)

between the lines representing the manual and non-manual social classes. In the non-manual the number of sibs was without effect in the vast majority of families, whereas in the manual class the effect began after the second child. Indeed, it seems clear at least in the UK, that the manual working group is itself heterogeneous. There is within it a subgroup smaller and more delayed to a disproportionate degree. In Rona's data for children aged five to 11 the difference between the 50th centiles for manual and non-manual was 2 cm, but the difference between their 3rd centiles was twice this. In the Newcastle-upon-Tyne study, reported in a series of elegant papers by Billewicz and his colleagues, the difference in age at menarche between manual and non-manual was 0.26 years; but this came entirely from that subgroup of manual workers who also had large families.

Association with unemployment   In England in the 1970s and 1980s there was a clear association between unemployment of the father (for over six months) and height of his children, and this, again, is much greater in the unskilled manual working class than in the others. The national surveillance system estimated the differences in length between children of classes IV and V in employment and not in employment as 0.8 cm and 0.6 cm, in boys and girls, at age one, and 2.0 cm and 1.5 cm at age two.

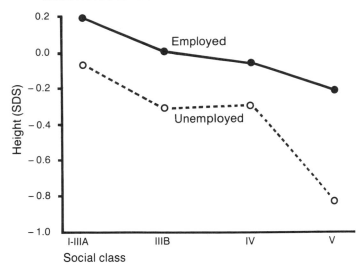

Figure 48   Height of 5–11-year-old children in England expressed as Standard Deviation Scores (SDS) according to social class and employment or unemployment of their fathers (From Rona *et al.*, 1978)

The relation at ages five to 11 is shown in Figure 48; the absolute difference in class V is about 3·0 cm (*see also* Wenlock *et al.*, 1986). Further analysis of the factors within the catch-all of social class that are of chief importance points to mothers' educational level (just as in the Third World), number of persons per room of accommodation, low income and total number of children in the family.

Association with parental size and social mobility   All the differences associated with father's occupation (or lack of it) may not be of direct environmental origin. Parental size contributes far more than any other factor to the variation of height amongst any group of children, and the different social classes, as we have seen, differ somewhat in adult height. Thus a part – it is statistically a small part – of the social class differences are associated with differences in parental size. These could of course be entirely caused by environmental factors, with that particular part of the parent-offspring correlation being due to a common poor environment. But they could also be caused wholly or partly in another way.

Social classes are to some extent endogamous and movement from one to another in some cultures seems to be linked with body size as well as with intellectual ability and physical energy. In Belgium, R. L. Cliquet (1968) has shown that young men who were entering an occupation

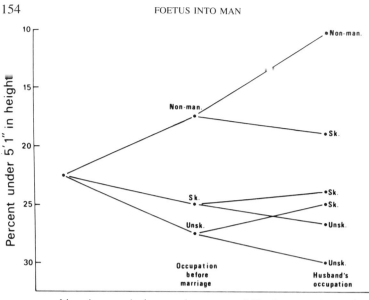

Aberdeen primiparae born to skilled manual workers

Figure 49 Percentage of daughters of skilled manual workers under 5 ft 1 inches tall taking non-manual and manual jobs and marrying men in non-manual, skilled manual and unskilled manual occupations
(Redrawn from Thomson, 1959)

exceeding in prestige that of their fathers were taller as well as cleverer than those who stayed in the same or an equivalent occupation. A similar social movement occurred amongst women in Aberdeen in the 1950s. Girls who entered non-manual jobs were taller, as usual, than those entering manual ones. In addition, however, the separation was increased in selection of a husband. The taller girls, both in the manual and the non-manual occupations, more often married men in non-manual jobs, and the shorter girls more often men in manual ones (Figure 49). Even amongst brothers, in a Polish sample investigated by Bielicki and Charzewski, the taller attained, on average, higher educational levels and had more highly-paid occupations.

Obliteration of the social class height differences   There are now a few societies in which the differences in height associated with occupation have actually been abolished. The first report of this was published in 1976 by Dr Gunilla Lindgren of the School of Education in Stockholm. In a survey of children in all urban areas of Sweden (though not, unfortunately, in rural ones) no difference according to father's occupation was found, either of height at a given age (between 7 and 17) or in age at menarche. In agreement with this, no difference in height according to father's occupation can now be demonstrated in Swedish men conscripted for military service. Health problems of preschool children in Sweden are no longer more frequent in the manual occupation class, as they are in most countries (Sundelin and Vuille, 1979).

Also in Norway; Dr Gro-Harlem Brundtland and her colleagues (1980) found no difference in 1975 in children coming from schools in Oslo districts with a high proportion of manual workers and those with a low

Figure 50   The mean height and weight of girls born in 1956 and measured in 1970 from 23 different schools in Oslo. The results are plotted against the percentage of manual workers in the corresponding electoral district in 1953. The lines are fitted by weighted least squares
(From Brundtland *et al.*, 1980)

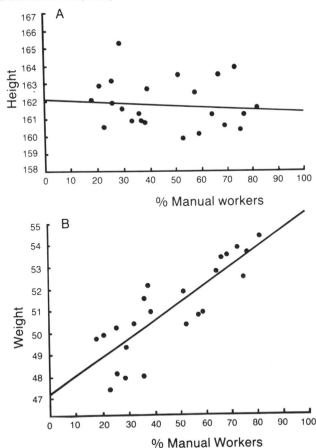

proportion. Figure 50 shows the data for just one sex-age group, 14-year-old girls. The regression of height on percent in manual occupations is not different from zero. At the time Dr Brundtland was Chief School Medical Officer of Oslo city; at the time of writing she is Prime Minister. Height is an accurate proxy for social conditions, that is when differences between groups are considered. Indeed, childrens growth provides a measure of one important aspect of the classlessness of society, and provides it in objective not rhetorical terms.

Height and mental ability There is a curious relation between height and measurements of intellectual ability. In all countries students are uniformly the tallest group in the population, averaging 2 to 3 cm taller than the mean of the total population (*see* Eveleth and Tanner, 1976). At the other end of the scale children in the UK National Child Development Survey designated as requiring placement in schools for the educationally subnormal were 3 cm less tall at age 11 than others, even when matched for father's occupation. Those who were in ordinary schools but who teachers thought might benefit from such education were about 2·5 cm less tall, and those receiving special help within normal schools were 1·5 cm less tall, again computing the differences within occupational groups of fathers (Peckham *et al.*, 1977). Part of this difference may be due to differences in tempo, but it is a common finding that mentally-handicapped adults are on average smaller than others. Why this should be so is not at all clear.

In schools there is a small correlation uniformly found between height and scores on tests of ability or intelligence, of the order of 0·3. Part of this is due to linked tempo differences in bodily and mental development. But part is not, for surprisingly the correlation seems to persist into adult life. In Swedish conscripts a correlation of 0·24 was found, and other data give similar values. Obviously such a small correlation is meaningless in the individual case, but it should tell us something about our culture. The situation is statistically comparable to accident rates, which do not tell an individual whether he personally will be killed at a particular cross-roads, but which nevertheless reflect accurately the state of the roads at that point.

9. Secular Trend
During approximately the last hundred years in industrialized countries, and recently in some developing ones, children have been getting larger and growing to maturity more rapidly. This is known as the 'secular trend'

in growth. Its magnitude is such that in Europe, America and Japan it has dwarfed the differences between occupational groups.

Figure 51 shows Swedish data, which are the most comprehensive available for any country over the whole period. The differences in children's heights between 1883 and 1938 amount to nearly 1·5 years growth; differences between1938 and 1968 are much less. At the age when growth ceases, as shown by the 16–18 year-old girls, the secular trend is still present, but much smaller than in childhood. A more detailed look is provided by Brundtland's Oslo statistics (*see* Figure 52). The increase in height at all ages was interrupted during the years of Nazi occupation in World War II, but later resumed.

The occurrence of a secular increase in height and weight has been documented in nearly all European countries, including Sweden, Finland, Norway, France, the UK, Italy, Germany, Czechoslovakia, Poland, Hungary, Soviet Union, Holland, Belgium, Switzerland and Austria. From about 1900 to the present, children in average economic circumstances have increased in height at ages 5 to 7 years by about 1–2 cm per decade. The trend starts early in childhood, as preschool data make clear.

Figure 51   Secular trend in growth of height. Swedish boys and girls measured in 1883, 1938–9 and 1965–71
(From Ljung, *et al.*, 1974)

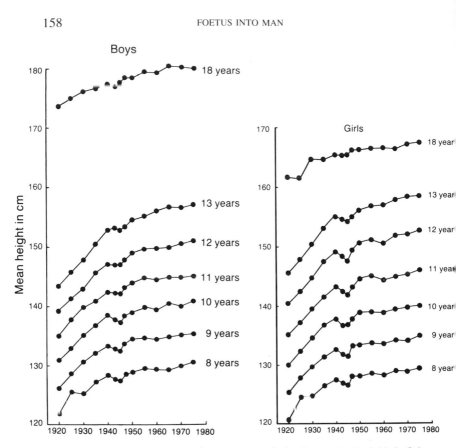

Figure 52   Mean height of successive age groups of schoolboys and schoolgirls in Oslo, 1920–1975. The age specifications refer to age group centres. The ages 8 to 13 cover the total population of schoolchildren in Oslo
(From Brundtland *et al.*, 1980)

At least in Britain it began a considerable time ago. In 1833, the first relatively large-scale measurement of children took place to provide parliamentary evidence on the effect of the employment of children in factories. At that time, working boys aged 10 years averaged 121 cm in height compared with 140 cm today; those aged 18 years averaged 160 cm compared with 175 cm today. These differences are actually a good deal larger than the differences seen at present between urban slum-dwelling children in the underdeveloped countries and the affluent children of the industrialized West.

The trend in Canada, the USA, Australia and other countries has been similar. Japan had a particularly dramatic increase in the years after the end of World War II. From 1900 to 1940 the trend for height was less than 1 cm/decade; but from 1950 to 1977 it amounted to 3 cm/decade in 7- year olds and 5 cm/decade in 12-year olds. In statistics which sample the whole population, Dr Hayashi of Kyoto, in association with the writer, showed that the height change in fully grown young adults was 2·8 cm in men from 1957–67 and 1·5 cm from 1967–1977. In women the figures were 1·5 and 1·2 cm. Astonishingly, sitting heights as well as stature have been for many years routinely recorded in Japanese children and young people, and the figures showed that the whole of this secular trend since 1957 was accounted for by changes in leg length: sitting height had not changed at all. Thus in the proportion of legs to trunk Japanese are now quite similar to Europeans, though their height on average is at about the European 15th centile.

In the industrialized countries the trend is now gradually stopping, as may be seen in the Swedish data. A particularly sophisticated calculation has been made from the surveillance data of the British National Study of Health and Growth on some 29,000 children aged five to 11 from 1972 to 1980 (Chinn and Rona, 1984). The trends were 0·8 cm/decade for English and 1·6 cm/decade for Scottish boys, and 0·5 cm/decade for English and 0·6 cm/decade for Scottish girls. Dr Albert Damon, of Harvard University, showed that amongst American families with sons educated at Harvard in successive generations the increase in height was 2·6 cm between the first two generations recorded (mean birthdates 1858 and 1888); 1·1 cm between the second pair of generations (mean birthdates 1888 and 1918) and nil between the third pair of generations (mean birthdates 1918 and 1941). In poorer sections of the community, however, the trend continues.

During the period since 1880 the upward trend in adult height in Europe has been about 1 cm /decade. An astonishing series of Norwegian growth data stretching back to 1741 reported by V. Kiil indicated little if any increase in male height from 1790 to 1830, and these and other data show a trend of some 0·3 cm per decade in several countries from 1830 to 1880, depending on their situation *vis-à-vis* the progress of the industrial revolution. As Dr Van Wieringen of Utrecht University has shown in analysing the data on height of Dutch conscripts from 1851 to the present, periods of economic setback were associated with a stoppage or even a reversal of the secular trend. At all times, however, the trend has been much smaller in adults than in children. Thus, much of the trend in children's heights is due to their maturing earlier. This is best shown by statistics on age at menarche, illustrated in Figure 53. The earlier data are all based on recollected age, and are therefore suspect in detail. However, the general trend is plain to see, and from 1880 to 1970 it averaged some three to four months per decade. This is not to say that the trend has always been as linear as this seems to imply, nor that it is the same for all social classes. The detailed study of Oslo Maternity Hospital records by Brudevoll and colleagues (Figure 54) shows a rather rapid decline from 1860 to 1890 followed by a period of little change until about 1920, when a rapid decline began again. This continued till about 1960, after when the decline gradually stopped. Most of the values plotted in Figure 54 refer to working- class women. Middle-class women in the period 1860–1900 had menarche on average one year to 18 months earlier. In the UK from 1910 to 1960 there was a drop of 4·4 months/decade in the working class, and 2·8 months/decade in the middle class. Wyshak's extensive data on middle-class white Americans 1925–55 give a drop of 3·2 months/decade. From 1965 to 1975 the trend was 3 months/decade in Polish cities, 7 months/decade in rural villages (Laska-Mierzejewska, 1982). During the last ten years the trend has slowed down or even stopped in most European countries.

Figure 53   Secular trends in age at menarche, 1860–1970
(Sources of data and method of plotting detailed in Marshall and Tanner, 1986)

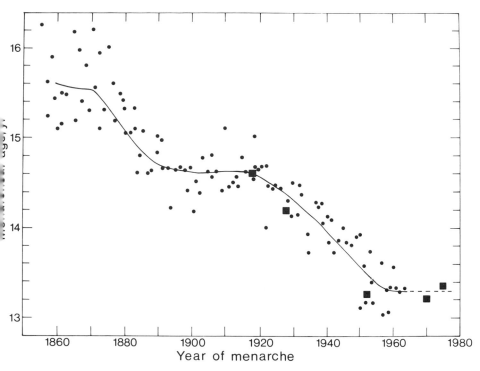

Figure 54   Mean menarcheal age for working-class women (1860–1945) and working-class and middle-class women (1945–1975) in Oslo, Norway. Recollection data: each point represents the average of about 50 maternity clinic patients. The squares represent *status quo* probit-fitted data on Oslo schoolgirls. The curve is a 21-term moving average until 1960 (Redrawn from Brudevoll *et al.*, 1979)

The time of the peak of the adolescent growth spurt has also become earlier, for example in Oslo 1920–1950 by 1·5 years in boys and 1·0 in girls, and in Japan 1957–1977 by one year in boys and 0·5 years in girls.

The causes of the trend are probably multiple. Certainly better nutrition is a factor, and perhaps in particular more protein and calories in early infancy. A lessening of disease must also have contributed. Ellison (1981) has shown that the trends with time in age at menarche and infant mortality amongst South American and Asian countries (the two groups considered separately) are strongly related. Estimates of nutritional intake are available over the same period, and these, in contrast, relate not at all to the change in growth tempo. On a between-population basis, then, we are seeing the same as we saw within populations: hygiene and delivery of medical care are of paramount importance. Some authors have put forward the rather fanciful idea that increased psychosexual stimulation consequent on modern urban living has contributed, but there is no positive evidence for this. Girls in single-sex schools have been studiously compared with girls in co-educational schools in Finland and Sweden, with totally negative results, though whether this is a fair test of differences in psychosexual stimulation may be open to debate. Climatic changes have also been suggested. It is true that the world mean surface temperature rose from 1910 to 1940, but since then it fell again, to reach 1910 levels by 1970 (Lamb, 1973). However, climate seems to exert at most a very minor effect.

An additional explanation has been suggested for the increase in adult size, though not for the increase in tempo. It may be that some degree of dominance occurs in some of the genes governing stature; this would mean that when a person carrying mostly genes making for smallness marries a person carrying mostly genes making for tallness the height of their offpsring would not lie on average exactly halfway between, but a little to the tall side. In more technical language, dominance is said to exist when some of the heterozygotes influencing stature produce effects nearer to the 'tall' than to the 'short' homozygotes. If this occurred (and only if) it would give rise to the phenomenon known as heterosis (or, in animals, hybrid vigour). An increased degree of outbreeding (i.e. of marriage to quite unrelated persons rather than cousins of various degrees) increases the number of heterozygotes, so if dominance did occur, then adult stature would be raised by the heterosis effect. There is increasing evidence that some degree of dominance in height genes, even if minor, may indeed exist. As for outbreeding, that has been steadily increasing ever since the introduction of the bicycle.

10.   Epidemiological Auxology: Height as a Proxy for Health

In much of what has gone before, we have been tacitly assuming that the greater the height (*not* the weight) the better. But this begs the question: does it matter to be small? Is small perhaps beautiful, here as elsewhere? Clearly we have to distinguish small as a – let us say – possession, and small as a symptom. A small man who has attained his genetic potential is one thing; a man who is small because he has not attained it is quite another. Can we demonstrate, in the latter case, that smallness is a measure of, or, as the economists say, a proxy for, health?

It seems at present that the only workable definition of health that we have, sadly, is absence of disease, a lower morbidity and mortality. The evidence relating shortness to morbidity has been briefly discussed above; but morbidity is hard to study accurately. As regards mortality, two studies may be cited. In the Gambian villages studied by Billewicz and McGregor of the Medical Research Council Unit, children who died before age five were significantly shorter (and lighter) than those who survived beyond age five. They were shorter at all ages, birth to death, so this was not a retardation of growth immediately before death, for those dying within three months of their last measurement were no smaller nor lighter than those dying six months or more after their last measurement. Shortness was clearly a proxy for tendency-to-die. How close the association is, and therefore how effective auxology is as a screening or predictive device, depends on the particular circumstances. In a town in Zaire, van Lerberghe (1988) showed the association was a good deal lower than in Gambia.

The second example comes from Norway. In the late 1960s all persons in Norway over age 15 were X-rayed in a mass miniature screening. At the same time heights and weights were recorded (except, unfortunately, in Oslo). About 1·8 million persons were examined, and when Waaler in 1984 reported on their follow-up, 176,000 had died. Shorter persons had a higher mortality at all ages until the over-70s. Amongst males aged 55–59, for instance, the mortality of those of height 185–189 cm (about 95th centile) was less than half the rate for those of height 165–170 cm (about 5th centile). During the first 12 years after the original measurement 12% of those at about the 25th centile for height died, as against 10.5% of those at the 95th. For the most part we are dealing with the generations who were children in the 1920s and 1930s, when Norway was less prosperous than now. Not only, it seems, is smallness in childhood a measure of health at the time; it may be a measure of the depth of life-long scars.

It does look, therefore, as though height can indeed be used as a proxy for health, and not only in the 19th century and in underdeveloped countries. This is true, of course, only when comparing *groups*, not when comparing individuals, the variation between whom is overwhelmingly due to genetical causes except in the direst circumstances. But between social classes, urban and rural dwellers, educated and uneducated (provided always we stay within a single ethnic group) height is a useful – perhaps the most useful – measure of healthiness.

Growth surveys and growth surveillance   For precisely this reason, many countries have made nationwide surveys of the growth of their children, and a few have instituted systems of continuous surveillance, designed to detect as early as possible a diminution of growth rate in any particular group – for example the unemployed in the population. The design of such surveys carries us beyond the scope of this book; for a single survey the example of the 1972 Cuban study (Jordan *et al.*, 1974) is outstanding; for the design and analysis of a continuous surveillance system the papers of Roberto Rona and Susan Chinn should be consulted.

Such surveys apply to groups. Using growth – or better *growth rate* – to monitor the health of individuals is another, and more difficult, matter. In the developed world, only very short-term or very often repeated and accurate measurements are likely to pick up the commoner ailments. In developing countries the situation is quite different, and following growth in length and weight, especially in infancy, may reveal individual problems of health and nutrition very effectively. For this the reader should consult a recent book edited by Waterlow (1988).

CHAPTER 10

The Organization of the Growth Process

We are now in a position to consider some basic principles that seem to characterize the growth process. We touched briefly on this topic in Chapter 2 when discussing cells as the basic unit of growth, but here we shall deal with the growth of the organism as a whole. Most of the examples are already familiar from earlier chapters.

Canalization and Catch-up

In Chapter 9 we saw that the birth lengths of monozygotic twins differed more than the birth lengths of dizygotic twins. However, by the time a few months had gone by, the reverse was the case. Those monozygotic twins who had suffered a poor position in the uterus had recovered from it and were back on their own trajectories of growth.

This is an example of what Waddington, in his classic book *The Strategy of the Genes*, called *canalization*. By this he meant that in any young animal growth has a tendency to return to its original path or channel if circumstances have conspired to push it off course. The processes of growth and differentiation are self-stabilizing or, to take another analogy, 'target-seeking'. The passage of a child along his growth curve can be thought of as analogous to the passage of a missile directed at a distant target. The target is determined by the genetic structure; and just as two missiles may follow slightly different paths but both end by hitting the target, so two children may have slightly different courses of growth but end up with almost the same physique.

The power to stabilize and return to a predetermined growth curve after being pushed, so to speak, off trajectory persists throughout the whole period of growth and is seen in the response of young animals to illness or starvation. During starvation an animal's growth slows down, but when feeding begins again its velocity increases to above normal for its age or maturity. Unless the starvation has been prolonged or has occurred very early in life, the original growth curve is caught up to and then once again followed. Figure 55 gives an example in a young baby who had two periods of starvation; other examples are shown in Figures 89 and 90 (pp. 230 and 234).

This rapid growth following the end of a period of growth restriction (for whatever reason) has been named *catch-up growth*. (It has also been

165

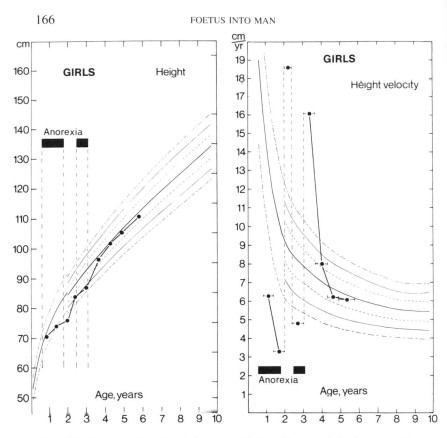

Figure 55   Length and length velocity curves of infant with two periods of undernutrition (anorexia) to show catch-up
(Redrawn from Prader, Tanner and von Harnack, 1963)

called 'compensatory growth' by animal nutritionists, but this phrase has been pre-empted by zoologists to designate in mammals the excessive growth of, for example, the second kidney when the first is damaged or removed, and in amphibians the re-growth of whole limbs following limb removal.) Catch-up growth may be complete and restore the situation, so far as we can tell, entirely to normal. There are two ways in which this may be done (Figure 56). In true complete catch-up the velocity increases to such an extent that the original curve is attained and thereafter growth proceeds normally (curve A in the figure). In complete catch-up with delay the catch-up velocity is not sufficient to do this, but maturity is delayed and growth is resumed at the correct velocity for chronological age (curve

Figure 56   Diagram of two sorts of complete catch-up. Growth retardation occurs from R to s with rehabilitation commencing at s. True complete catch-up A; complete catch-up by prolonged growth B and C. In C growth is resumed at average velocity for chronological age; in B at average velocity for bone age, marked RUS

C) or, more usually, for maturity as signified by bone age (curve B). In these cases growth continues for longer, but in the end complete catch-up, or compensation for the growth arrest, is achieved.

Frequently the catch-up is a mixture of these responses. In Figure 57 are shown the height curves of two brothers both of whom had isolated growth hormone deficiency from birth. The elder was not treated till age 6·2 years. By this time, besides being small, he was much delayed in growth, with a bone age of 3·0 'years'. He had a marked catch-up velocity, shown in Figure 58, but it was insufficient to bring him within the normal range of height-for-age curves. He remained somewhat delayed and his adolescent spurt took place some eighteen months later than average. Even so he ended only a little below the range of variation expected from his parents' heights (shown in the figure by the vertical thick line with м and ғ, representing mother's and father's centiles, upon it). His younger brother, whose treatment started much earlier, at age 2·1 years, showed more complete catch-up, reaching the 10th centile after 5 years of

Figure 57   Heights of two brothers with isolated growth-hormone deficiency treated with HGH from ages 6·2 years and 2·1 years. Catch-up of older brother is partly by high velocity and partly by prolonged growth and is incomplete. Younger brother shows more complete catch-up. ғ and м, parents' height centiles; vertical thick line, range of expected heights for family

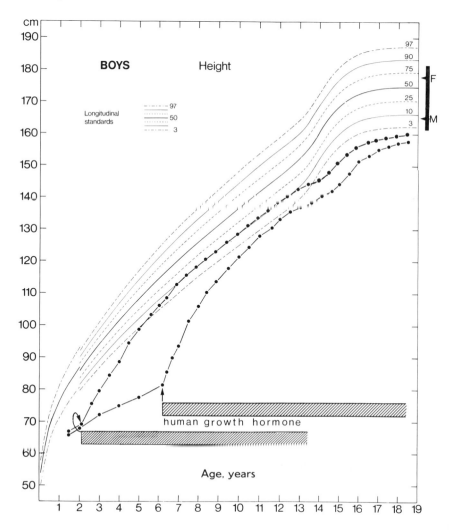

treatment and thereafter continuing along it (notice this is his mother's centile).

Children have an astonishing capacity to return to their growth curves, and a 5-year-old whose growth has been quite arrested for a whole year, e.g. by hypothyroidism (*see* Chapter 12, p. 234) will catch up completely provided he is well fed and looked after during his period of rehabilitation. Pathology or undernutrition early in life is another matter and children starved early in utero through some imperfection of the placenta usually

Figure 58 Height velocities of brothers shown in Fig. 57. Arrows signify beginning of HGH treatment

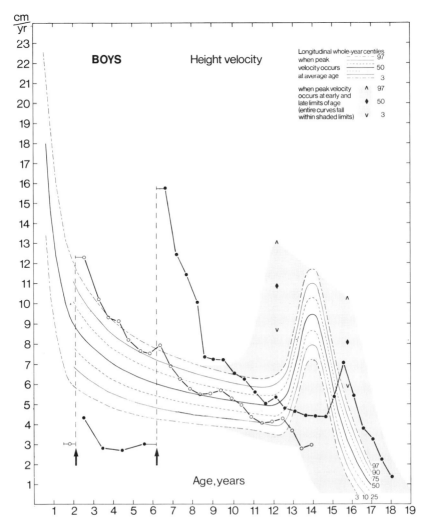

fail to catch up completely. We do not know for certain why this is so, but it is thought that it relates to the phase of cell multiplication. As explained in Chapter 2, it seems that cell division has to take place to a tightly organized schedule and cannot be delayed while the organism waits for better conditions, as can the later phase of packing material into the cells. An alternative, or more probably a complementary, explanation relates to one of the few extant theories of the catch-up mechanism. This is based on the fact that the child's catch-up velocity is large to start with and decreases progressively as the target curve is approached. It looks, therefore, as though the velocity is regulated by a signal of mismatch between the ideal or target curve and the actual situation. If so, both must be in some way monitored. The theory proposes that the ideal, genetically programmed, curve is represented in the brain (probably the hypothalamus) by a process or tally (perhaps the accumulation of specific substances), which continues almost regardless of outside events. The actual situation may be signalled by molecules 'spun off' each time a molecule of protein is synthesized in cells of muscles, cartilage or even liver, or perhaps when cell division occurs in the generative zone of the cartilage cells. These molecules would interact with the brain tally, cancelling it. The growth velocity would be proportional to the uncancelled amount of the tally.

This is a highly speculative scheme, possibly quite wrong and certainly incomplete. At least, however, it provides a framework for thought and experiment. It explains the difficulty of compensating for early prenatal growth failure by supposing that the brain process representing the genetic programme can be affected by lack of the proper supply of molecules at the time when the process is being constructed. This would be in early foetal and mid-foetal life. If it occurred, then the target itself would have been lowered and catch-up would henceforth be to this lowered target, instead of to the genetically programmed one. Later, when the neurons had finished forming or perhaps finished some aspect of their protein synthesis, the target would become more nearly inviolable.

Evidence in favour of this has recently been provided by Dr David Mosier at Irvine, California. If two-day-old rats are subjected to head irradiation of 600 rads (the pituitary gland being shielded), they grow up to be smaller than controls, secreting smaller amounts of growth hormone per burst, though no fewer bursts. If they are starved for 48 hours when aged 40 days, and then refed, they have a catch-up during the refeeding just as normal rats do. But this catch-up restores them not to the normal rat curve, but to the growth curve of their irradiated, but not starved, littermates. The central 'sizo-stat' seems to have been altered by the irradiation.

The proximate cause of the greater-than-usual velocity in catch-up is not known. Evidence is contradictory on whether increased secretion of growth hormone or somatomedin occurs. Some circulating messenger seems likely since all cells of the body appear to join in the catch-up spurt. Our ignorance about the catch-up mechanism is not too surprising when we reflect that we do not even know why younger animals grow faster than older ones. One thing does seem to be clear about canalization. Regulation is better in females than in males. This applies to Man as well as to other mammals. The radiation of the atomic bombs in Japan, for example, slowed down the growth of exposed boys more than that of girls, and the same is generally true of undernutrition except in social situations where the male is clearly favoured. Similarly, girls recover from growth arrest more quickly than boys. The physiological reason for this greater stability is not known.

Competence and Specification

All cells have a full set of chromosomes and hence of potential assembly-lines for all types of protein (see Chapter 2, p. 26). But very early in embryological development this general competence, as it is called, is lost; each cell concentrates more and more on certain specialized production-lines. It seems that the other lines are put into mothballs rather than actually smashed up, but the moth-balling is usually pretty thick. Thus when the period of competence of a cell to do something has passed, it is usually impossible to reactivate it.

The nervous system provides a good example of this property. The developmental programme of nerve cells is under the control of genes. At each stage of development some genes are turned on, and others turned off, usually for ever. The regulator molecules which do the turning on and off may either be resident in the cells themselves or may diffuse out of the cells in which they are produced and regulate neighbouring cells. Thus they may exert differential actions on cells at different distances away or on cells in different positions in the embryo. In mammalian and amphibian development the diffusing molecule mechanism is the chief one; a field of differential distribution of regulating signals is set up and cells are labelled according to their position in this field.

The most throughly studied instance is that of the toad retina. There is a small area of the embryonic nervous system which will develop into the neural part of the retina. It is polarized very early into latent front and back ends, as can be shown by taking it out of the body and growing it in tissue culture. However, if it is excised while still an early eye-rudiment and replaced back-to-front it is able to change its polarity and a normal eye is formed. The change is made through interaction with the rest of the system surrounding it. But if the rudiment is excised, left to develop to a more advanced stage outside the body in tissue culture, and then reimplanted back-to-front it can no longer reverse polarity and the eye forms an inverted pattern of nerve connections with the brain. This time the neurons are said to be 'specified' or 'determined'. Complete specification seems to accompany what Jacobson calls 'withdrawal from the mitotic cycle', i.e. cessation of the power of dividing. The time at which specification occurs varies from tissue to tissue and even from cell to cell. Specification usually occurs long before any morphological signs are visible; thus an experiment may reveal that certain cells have been specified irreversibly to make a certain part of the brain; but only considerably later can they be identified as actually turning into that part. Specification, in others words, is a matter of molecular form and distribution. The effect of a central-nervous-system poison in the foetus (e.g. thalidomide) depends firstly on the specific chemical affinities of the poison with particular types of cell and secondly on whether these cells and their potential replacements are specified or not at the time when the poison acts.

Sensitive Periods

A very important extension of this principle is seen in the concept of *sensitive periods*. They were originally called 'critical periods' both by biologists and psychologists, and have also, in a more limited context, been referred to as 'periods of special vulnerability'. We have already met several examples: the period when, just after it begins to use its eyes, the kitten must receive the impact of the visual environment (p.116); the immediately neonatal period when the male rat's brain must receive testosterone from the animal's testes (p.56); and the period in the 14th postmenstrual week when the human male foetus must receive male sex hormone for differentiation of the external reproductive organs (p.54).

All these are critical periods in the ordinary sense of the word, but in the physiological sense they are periods of increasing sensitivity of a receptor to a highly specific stimulus, followed by decreasing sensitivity and eventually by total lack of response. In the case of the kitten, for example, closure of eyes for 3 or 4 days at the height of the sensitive period in the 4th week after birth gives the same effect as closure for a month at

8 weeks. In the case of the rat's brain-differentiation the sensitivity can be accurately measured by seeing what dose of administered testosterone suffices to bring about the effect in the castrated rat. No amount will do the job before birth; a large amount is necessary during the first 2 days after birth, and a much smaller amount during the subsequent 3 days. Then after about 2 further days the period is past and once again no amount of testosterone can make good the damage.

There is a familiar example in human pathology, that of rubella, or German measles, in the pregnant mother. Some children whose mothers have rubella between the 1st and 12th weeks of pregnancy are born with cataracts in the eyes, and other defects. Rubella after the 12th week has no effect, so the period is a critical one of vulnerability.

The physiological basis of sensitive periods is not known in detail, but the general outline is clear enough. In the case of the rat there must be either a rising and waning amount of testosterone receptor in the particular part of the brain concerned, or else a rising and waning affinity of receptor to hormone. What is not clear is what causes the timing of this transient wave of sensitivity. If we knew this we should know much about the process of growth, for it seems as if sensitive periods are universal in all stages, at least of early development. One wave forms, rises, falls; its fall sets off another and *its* fall, a third.

Sensitive periods seem to be less frequent during postnatal growth, although this may merely reflect our inability to detect them. In the Rhesus monkey, Harwerth and others at Houston have recently mapped the sensitive periods for various visual functions. By three months after birth the period for retinal rod function is closed (i.e. covering one eye for longer results in a permanent deficit). The period for retinal cone function extends to six months; that for spatial vision, or detection of form, till 25 months and that for binocular vision for some period in excess of 25 months. There are clearly multiple, partially overlapping sensitive periods for different functions. Very roughly, these ages in Rhesus can be converted to the equivalent in the human by multiplying by four. If similar periods occur for other brain functions, they would be of the greatest concern for educationists. It is widely believed, on educational grounds, that periods of sensitivity to various environmental sensory inputs indeed exist. Certainly there is a formal analogy between 'readiness to read', for example, and competence of brain systems. It is possible that the connection will be shown to be more than formal when research on brain growth has sufficiently advanced.

Heterochrony and Disharmonic Development

The multitude of chemical reactions going on during differentiation and growth demands the greatest precision in linkage. Thus for normal acuity of vision to occur the growth of the lens of the eye has to be harmonized closely with the growth in depth of the eyeball, so that the point of focus of the light-rays lies exactly on the retina. It is small wonder that the success of this co-ordination varies and most people are just a little far-sighted or near-sighted. Again, it seems that many features of the face and skull are individually governed by genes which do not much influence the growth of other, near-by features. But in general the parts of the face fuse to form an acceptable whole. In this case the final growth stages are plastic and variable, and in fitting together, for example, upper and lower jaws, forces of mutual regulation come into play which do not reflect the original growth curves of the discrete parts.

The regulative forces harmonizing the velocity of growth of one part with that of another do not always succeed. If the original genetic forces begin by being too unbalanced, as in Down syndrome (*see* p. 54), normal development cannot occur.

Variation in the speed of development of different structures and functions (heterochronisms) underlie many individual differences in bodily structure. Examples are the longer arms and legs relative to trunk in men as opposed to women, or in Africans as opposed to Europeans.

This general point of view, derived from experimental embryology, clearly has possibilities for behavioural research, even though it would be unwise to press the analogy too far. Not all human behaviour is open to such delightfully simple explanations as that of *Anser oedipoidipus* studied by Konrad Lorenz. Domesticated strains of geese mature earlier than wild geese, and in the offspring of a wild gander and a domesticated goose disharmonization occurs between sexual maturation and the earlier mother-following response. The mother-following response still remains operative when the sexual response appears, and the young male bird consequently insists on copulating with its mother. Since the wild father's sexual activity arises only later in the spring, it is unnecessary that he be killed first; he remains wholly indifferent to the drama.

The behaviour of children with precocious puberty provides an instructive example of an extreme, pathological degree of heterochrony, in which a fully developed endocrine system acts upon a less developed brain. Psychosexual advancement by no means keeps pace with endocrine development. The hormones need a mature brain equipped with adolescent experience to work on if adult sex behaviour is to occur. This is not to say that the hormones are without effect. In one of the best described cases to date, by Drs Money and Hampson of the Johns Hopkins Hospital Psychoendocrine Clinic, the psychosexual development of a boy with simple precocious puberty was at a level characteristic of his chronological age, but more energized than is normal. The boy was 6·5 years old, with a bone age of 15. He had begun to have seminal emissions at 5 and to masturbate at 6. He experienced numerous dreams involving kissing women all over the body, which he would relate only to a male interviewer, with the air of a 'roué narrating his escapades in all-male company'. He had no knowledge of copulation, and overt sexual behaviour towards women had never been a problem. Several women on the hospital staff, however, said they felt uncomfortable under his gaze, which carried a considerable message of seduction.

Growth Maturity Gradients

Onc way in which the organization of growth shows itself is through the presence of maturity gradients. One such is illustrated in Figure 59. In Figure 59b the percentage of the adult value at each age is plotted for foot length, calf length, and thigh length in boys. At all ages the foot is nearer

Figure 59 Maturity gradients in length of: (*a*) upper limb; (*b*) lower limb (From Tanner, 1962)

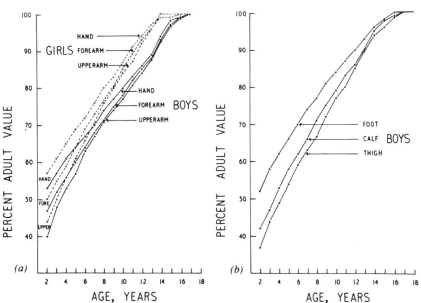

its adult status than is the calf, and the calf is nearer than the thigh. A maturity gradient is said to exist in the leg, running from advanced maturity at the far end of the limb to retarded maturity at the end nearest the trunk. The word 'gradient' has arisen because of the supposed mechanism by which this occurrence takes place. It is thought that in the embryonic limb bud, before any differences between the three segments can be discerned, there must be differences in the concentration of some chemical substance. Thus a concentration gradient of the chemical substance leads eventually to a maturity gradient in developing physical structure.

Figure 59a illustrates the same gradient in the arm, together with the fact that girls are more advanced towards maturity than are boys, without this affecting the gradient within the limb.

Many other growth maturity gradients are known, some covering small areas only and operating for short periods, other covering whole systems and operating throughout the whole of growth. The head is at all ages in advance of the trunk, and the trunk in advance of the limbs. In the hand and foot the second digit is the most advanced, and the third, fourth and fifth follow in that order.

Much of the growth of the brain is organized by means of such gradients and several have been described in Chapter 8.

Stages of Development: General and Singular

In child development there has been much argument as to whether development is continuous or whether it occurs in stages. Again, supposing stages do exist, there is argument as to whether any *general* ones occur, characterizing a simultaneous achievement of a number of anatomical, physiological and psychological developments.

In physical growth, there is continuity and little evidence for discrete stages. Even in the development of motor skills such as crawling and walking, development is more continuous than sudden. It is sometimes said that a child 'walks or does not walk', in a tone which implies that one day the child, previously unable to walk at all, suddenly finds the ability to walk unaided. Close observation of the development of such an ability, however, by no means supports this notion; on the contrary, the ability to walk develops gradually, as Dr Myrtle McGraw carefully documented many years ago.

In some creatures general stages of development do occur; e.g. in insects at metamorphosis. But in man only at adolescence do we have anything approaching a general stage; here, it is true, developments in anatomy, physiology and behaviour tend to occur simulaneously. Even here, however, the degree of synchrony is only relative. Skeletal age and dental age are largely independent of each other; and the spurt in bodily development is not matched by any great rise in intellectual capacity.

Physical development is best envisaged as a series of many successive processes, overlapping one another in time and linked loosely or tightly as the case may be. Out of the complexity of the linkages, under equilibratory forces, emerges an overall order in which visible changes follow one another with the regularity of a serially changing mosaic. The process is one of continuous unfolding and movement, with speeds varying from time to time in different parts of the mosaic; it is not a series of kaleidoscopic bumps. Only in certain restricted areas do rapid reassortments of the pieces occur, as they fall into new and increasingly precise patterns.

CHAPTER 11

Standards of Normal Growth

This chapter brings together, in a usable form, standards for normal growth and growth velocity in height and weight. It is a do-it-yourself kit for parents who want to know how tall their children will grow to be and for teachers or health visitors who want to see whether a child in their care is growing properly or not. As in all cases of do-it-yourself, the precision of the job depends on the care and understanding with which it is done. There is no inherent difficulty, except where X-rays are concerned.

Two sets of standards are presented here; one for British and the other for North American children. The British standards are derived from children measured in the 1960s. For school-age children (5–16 years) a random sample of London schools was selected, and all pupils in them were measured, ensuring approximately 1,000 boys and 1,000 girls for each year of age. The preschool sample was derived from all births registered in a central area of London but was much smaller, as tends to be the case in all longitudinal surveys. However, a recent (1975–85) large-scale nationwide survey of English preschool and school children (*see* Rona and Altman, 1977) shows these standards represent the present situation in the UK very well.

One or the other of these standards should be fairly adequate for judging the growth progress of any child of European or African origin in a developed and relatively temperate country (though *see* p. 209; several countries have recently developed similar standards of their own). With suitable caution they should apply also to children growing up under good social, educational and nutritional conditions in many developing countries. Children of Far Eastern origin will be slightly smaller and have their peak velocities about a year earlier (*see* Chapter 9). All the same, the reader who wishes to use these standards should be careful of two things. First and foremost, the lengths and heights *must* be taken with proper care ('garbage in, garbage out' is the auxologist's motto no less than the computer scientist's). Second, the various factors affecting growth described in Chapter 9 should be borne in mind in interpreting the results.

Collections of growth data may be used in three distinct ways. First, growth standards may be constructed to serve as a screening device – a very powerful one – for investigating groups of not overtly ill children to see which individuals might benefit from special medical, educational or social care (the *screening* or *community care* use). Secondly, growth standards may be constructed for studying the response to treatment of a child known to be ill (the *paediatric* use). Thirdly, growth data can be collected, and used as an index of the general health and nutrition of a population or sub-population as described in Chapter 9. Such data can then be compared with simultaneously collected data from other populations or sub-populations (surveys) or with data collected on the same population or sub-population after the passage of some years (surveillance) or after some remedial, social or economic action. The population may be that of a whole country, or of a particular ethnic, social or occupational group within a country. For this use standards as such are not necessary; comparisons proceed by the classical statistical methods. In this chapter we are concerned only with the first two uses of growth data.

The best way to begin is by measuring and plotting (as if at age 19) the heights of father and mother and any grown-up brothers and sisters. The mother's height cannot be plotted unadjusted on the boy's chart or she will look ridiculously small. First the average difference between men and women should be added, which is approximately 13 cm (5 ins). In this way the mother's height *centile* rather than her absolute height will be plotted. Similarly 13 cm must be subtracted from the father's height before plotting it on a girls chart. These plots done, the mid-point of the father's and mother's centiles is found and a line is drawn upwards from it for 10 cm in a boy's chart, 9 cm in a girl's. A corresponding line is drawn downwards for 10 cm or 9 cm. The band thus formed is known as the 'parental target range'. It gives the range of height centiles within which 95% of the sons or daughters of the couple will be expected to fall as adults (Figure 60). (Strictly speaking it is the centile of the mid-parent height [average of father's and mother's heights] which should be plotted but the procedure described gives an acceptable approximation.) If the average parental centile is the 25th, the expected or 'target' range runs from below the population 3rd to about the population 75th centile. Such parents should not therefore expect all their children to be at the 50th centile, nor any to be at the 97th. Note that a particular child's expected height may lie anywhere within the target range and should not be supposed to be at the central bulls-eye. All the centre of the range indicates is the expected average of a large number of siblings. Standards which specifically allow for parents' height appear later in this chapter (p. 211).

Figures 60 and 61 show the standards for height for boys and girls. These are longitudinal-type standards as explained on p. 10, in which an average child having his growth spurt at the average time does actually

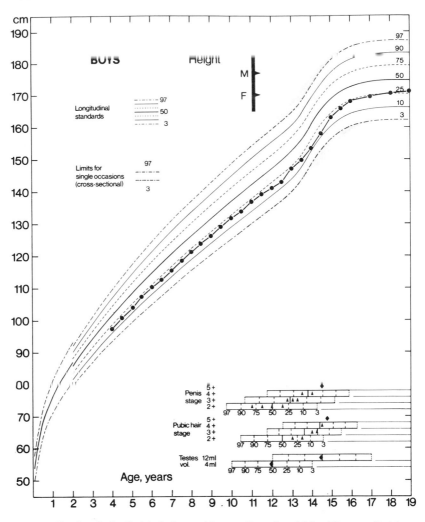

Figure 60   Standards of height for boys, with normal boy plotted. M and F, parents' height centiles; vertical bar, range of expected heights of offspring of these parents

follow the 50th centile. In Figures 60 and 61 examples are shown, with parents' centiles plotted. The open circles plotted in Figure 61 represent the height plotted against the *bone age* instead of chronological age, so the length of the lines from closed to open circles represents difference between chronological age and bone age. Bone age is a specialized

Figure 61 Standards of height for girls, with normal girl plotted (solid circles). Open circles plot RUS bone age. M and F, parents' height centiles; vertical bar, range of expected heights of offspring of these parents. Puberty ratings as shown

investigation needing medical consultation (*see* Chapter 6, p. 76). We shall consider its use later. The boxes in the lower right-hand portion of the charts relate to puberty standards and are also discussed later.

Decimal age   The age scale has been calibrated in tenths of a year, not in months. This is designed to make things simpler, once the initial shock has worn off. When we calculate growth *velocity* we find it is straightforward if we use years and decimals of years, but horribly complicated (if done accurately) using years and months. Accordingly Table 4 gives the decimal calendar. The year is divided into thousandths and each date has a decimal number; thus 1 May 1975 is 75.329. The child's birthdate is similarly recorded; a child born on 23 June 1969 has the birthdate 69.474. Age at measurement is then obtained by simple subtraction, e.g. 75.329 – 69.474 = 5.855, shortened to 5.86 years. This is the value read on the age axis in plotting height for age. Suppose the next examination is done on 15 March 1976. This is the date 76.200; the child is 76.200 – 69.474 = 6.73, and the increment period is 6.73 – 5.86 = 0.87 years. Suppose the child has grown 4.9 cm in this time. His growth rate would then be 4.9 ÷ 0.87 = 5.63 cm/year and the plot would be made in the velocity chart (*see* p.188) at the centre of the age interval, i.e (6.73 + 5.86) ÷ 2 = 6.30 years.

Measuring Techniques

The measuring technique is all-important, especially if velocities are going to be considered, since an error in the direction of low values on the second occasion can actually have the child growing down-wards. If the proper instruments and technique, shown in Figures 62 and 63, are unavailable, standing height can perfectly well be measured against the traditional door, with a triangular or bookshaped object brought down on to the child's head and the door marked. It takes at least two people to do this, however. One has to position the child, and another, at the right moment, brings the book, or, better, an architect's drawing triangle, down on the child's head. The longer side of the book (the spine) or of the triangle should be in contact with the door, the shorter with the top of the head. The side in contact with the head must, however, extend well out, at least to a point above the forehead. The child stands with his heels against the bottom of the door; his shoulders, buttocks and head may touch the door, but that is not necessary. He looks straight forward (not nose-in-air) so that the earholes are in the same horizontal plane as the lower border of the eye sockets. He is told to reach up as much as he can, make himself tall, take a deep breath and relax his shoulders, and while this is going on the observer applies gentle pressure upwards to the bony prominences just behind the ears. The heels must not be raised from the ground and another observer may be necessary to ensure this. When the exact position has

Figure 62 Measurement of height (Courtesy of R. H. Whitehouse)

been reached the triangle is brought down on top of the head and the position of its corner marked on the door.

It will be appreciated how useful a counterbalanced headboard resting at all times on the child's head is, and how useful the direct digital read-out of the modern machines, the design of which we owe to the ingenuity of R. H. Whitehouse. The machine enables a single observer to take the measurement, provided only that he can trust the child not to lift his heels. Wavy metal rods attached to the backs of weighing-machines, or pieces of wood moving in ill-fitting metal collars, are hopelessly inaccurate and should be rejected in favour of the door-and-triangle approach.

It is much harder to find a home-made substitute for the supine length machine. The best that can be advised is a right-angled board, such as a bed with a low headboard and a plank put along it. the infant's head must be held as shown in Figure 63 with the ear-eye plane this time vertical. The ankles are gently pulled to stretch the child, the legs are straightened, the feet turned up vertically, the triangle brought against them and its position on the bed marked. Supine length averages about 1 cm longer than standing height, but this difference varies from one child to another, its range running from 0 to 3 cm. In clinical work supine length is taken till age 2·0 years, and standing height thereafter, and the charts are constructed in this sense. From 2·0 to 3·0 years, however, both measurements are done, partly so that the difference can be evaluated and partly to ensure true

TABLE ~ *Tables of Decimals of Year*

	1 JAN	2 FEB	3 MAR	4 APR	5 MAY	6 JUNE	7 JULY	8 AUG	9 SEPT	0 OCT	11 NOV	12 DEC
1	000	085	162	247	329	414	496	581	666	748	833	915
2	003	088	164	249	332	416	499	584	668	751	836	918
3	005	090	167	252	334	419	501	586	671	753	838	921
4	008	093	170	255	337	422	504	589	674	756	841	923
5	011	096	173	258	340	425	507	592	677	759	844	926
6	014	099	175	260	342	427	510	595	679	762	847	929
7	016	101	178	263	345	430	512	597	682	764	849	932
8	019	104	181	266	348	433	515	600	685	767	852	934
9	022	107	184	268	351	436	518	603	688	770	855	937
10	025	110	186	271	353	438	521	605	690	773	858	940
11	027	112	189	274	356	441	523	608	693	775	860	942
12	030	115	192	277	359	444	526	611	696	778	863	945
13	033	118	195	279	362	447	529	614	699	781	866	948
14	036	121	197	282	364	449	532	616	701	784	868	951
15	038	123	200	285	367	452	534	619	704	786	871	953

	JAN 1	FEB 2	MAR 3	APR 4	MAY 5	JUNE 6	JULY 7	AUG 8	SEPT 9	OCT 10	NOV 11	DEC 12
16	041	126	203	288	370	455	537	622	707	789	874	956
17	044	129	205	290	373	458	540	625	710	792	877	959
18	047	132	208	293	375	460	542	627	712	795	879	962
19	049	134	211	296	378	463	545	630	715	797	882	964
20	052	137	214	299	381	466	548	633	718	800	885	967
21	055	140	216	301	384	468	551	636	721	803	888	970
22	058	142	219	304	386	471	553	638	723	805	890	973
23	060	145	222	307	389	474	556	641	726	808	893	975
24	063	148	225	310	392	477	559	644	729	811	896	978
25	066	151	227	312	395	479	562	647	731	814	899	981
26	068	153	230	315	397	482	564	649	734	816	901	984
27	071	156	233	318	400	485	567	652	737	819	904	986
28	074	159	236	321	403	488	570	655	740	822	907	989
29	077		238	323	405	490	573	658	742	825	910	992
30	079		241	326	408	493	575	660	745	827	912	995
31	082		244		411		578	663		830		997

Figure 63   Measurement of supine length (Courtesy of R. H. Whitehouse)

velocities over this period, with each measurement done the same way on the two successive occasions.

It was discovered over two centuries ago (in the context of advising destitute or adventurous young men how to appear at their tallest when applying to join the army) that measured height decreases during the course of the day. The decrease is due entirely to changes in the spine, mostly postural, and the technique of measurement described above minimizes these. Even so, some shrinkage remains, most of it during the first half-hour after rising from bed. Thus stature should ideally be measured just after breakfast. Even in the afternoon half-an-hour's horizontal rest will restore the situation to near the morning one; otherwise measurements in the late afternoon are apt to be up to 1 cm less than in the morning.

Interpretation of Plot

The centiles shown in the standards have been explained in Chapter 1, p.19. Height is Normally distributed, so the 50th centile corresponds to the mean of the population. The 3rd and 97th centiles, below and above which 3 in every 100 normal children lie, are the conventional limits of normality, or, more truthfully, of suspicion. We have to remind ourselves that 3% is not a negligible number. If every child under the 3rd centile in London came to the Growth Disorder Clinic at the Hospital for Sick Children then Great Ormond Street would be as crowded as Lourdes. Really pathological children lie a good way below the 3rd centile line.

Longitudinal versus Cross-sectional Standards

The standards in Figures 60 and 61 represent the curves of children who have their adolescent spurts at the average time. If a child is of average length at birth, has an average velocity throughout childhood, has the peak of his adolescent height spurt at the average time, ceases growth at the average age for doing this and ends up at the average height, then he will follow the 50th centile line exactly. But if he has an early adolescent growth spurt his curve will leave the 50th to rise above it, as shown in Figures 67 and 68 below. Since on average the final adult height of early-middle- and late-maturers is the same, the early-maturer's curve will descend again, ending at the same 50th as the average-maturers. If the boy is 50th centile for height but a late maturer, then his curve will fall below the 50th temporarily and later catch up to it. Early and late maturers should be judged, of course, according to their own curves and not the curves of the median-maturing cohort. Thus these standards are called, technically, 'tempo-conditional', meaning standards which allow for differences in tempo between children. The older cross-sectional, tempo-unconditional standards do not do this and are both biased and inefficient for studying the growth of individuals after the age of 9 years. They also show a wider variation for the reason discussed in Chapter 1, p.20. The 3rd and 97th centile of such standards are represented in Figures 60 and 61 by the outside of the shaded areas. It is logical to regard them as the limits of normality when a child is seen for the first time; but if the child's plot does fall in the shaded area, then by the second occasion of measurement, if he is normal, it will have moved towards the corresponding centile of the longitudinal standards, i.e. to within the non-shaded centile area.

Velocity Standards

More important than the child's exact position in relation to the 3rd centile in the distance chart is his plot in the chart of velocity. Most children with growth problems have a low velocity of growth. Velocity picks out the pathological cases better than does distance (i.e. height for age) because velocity represents what is happening *now*, whereas distance represents the sum of all that has happened in the past. Thus a child who at age 6·0 years was at the 50th centile for distance and failed entirely to grow for a year would still be well within normal limits for distance at age 7·0. A study of his velocity from 6 to 7 would at once show him to be totally abnormal.

Furthermore, the normality or otherwise of velocity from one age to the next cannot be properly judged by calculating the amount by which the

Figure 64   Standards for height velocity in boys, with normal boy of Fig. 60 plotted over whole-year periods, with new period commenced every 6 months
(From Tanner and Whitehouse, 1976)

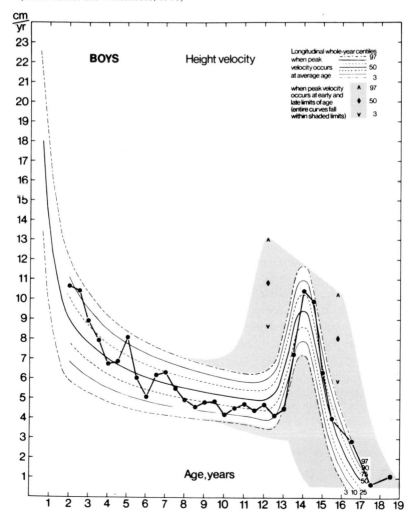

two distance scores – whether in centiles or standard deviations – differ. A change from -1SD to -2SD of height attained does not represent the same centile of velocity at age three, for example, as at age twelve. Though it is perhaps not intuitively obvious, the change of distance measure is biased and give a somewhat erroneous result; it is better to use the velocity.

Figures 64 and 65 show the British velocity standards for boys and girls. These are whole-year velocity standards, that is they represent

Figure 65 Standards for height velocity in girls, with a normal girl plotted over whole-year periods, with new period commenced every 6 months (From Tanner and Whitehouse, 1976)

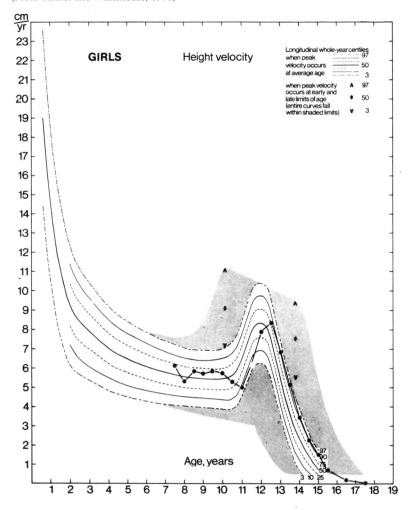

velocities not only computed in centimetres per year, but measured over approximately a year also. If a shorter interval is used then the variation amongst normals is greater. This is partly because the two errors of measurement assume a relatively greater importance in comparison with the actual growth that has occurred, and partly because of the seasonal influence making growth usually faster in the spring. For judging shorter periods than a year, therefore, the centiles should be farther apart. The children represented in Figures 64 and 65 are plotted over whole-year periods, but a new yearly period is begun each 6 months, representing the sort of rolling average much used by economists (e.g. when the yearly rate of inflation is computed at the end of each month).

The velocity centiles at adolescence represent the situation for that cohort of children all of whom have their peak height velocity at the average age for peak height velocity; i.e. 12·0 years in girls and 14·0 years in boys. This is the average-tempo cohort of children, and such children will follow one of the centiles during the spurt. In these charts the shaded portions represent the situation for early and late maturers. Boys who have their peak height velocity 1·8 years early (1·8 years being 2 standard deviations of 'earliness' or the equivalent of the 3rd or 97th centile) have higher peaks and are represented by the diamond (50th centile) and the two arrows (3rd and 97th centile) at 12·2 years. Boys with a similarly late spurt are represented by the marks at age 15·8. All normal boys' velocity curves should fall within the shaded area, and have a shape approximating that of the centile lines between the areas. Curves for an early- and a late maturing boy are shown in Figure 66.

One point of warning about velocity curves must be given. In distance curves a child tends to stay in the same centile throughout his whole growth, on average. But in velocity he cannot do so, unless his velocity is actually at the 50th centile. A child whose velocity is always at the 10th centile ends up pathologically small; and one whose velocity is consistently at the 80th ends up pathologically tall. Thus the pattern of an individual's velocity points throughout childhood is not the same as the pattern of his distance points.

Standards For North American Children

On average, children in North America have a slightly faster tempo of growth and grow to a slightly greater adult height than British children. Longitudinal or, as they are technically called 'tempo-conditional', standards for these children have therefore been developed (*see* Tanner

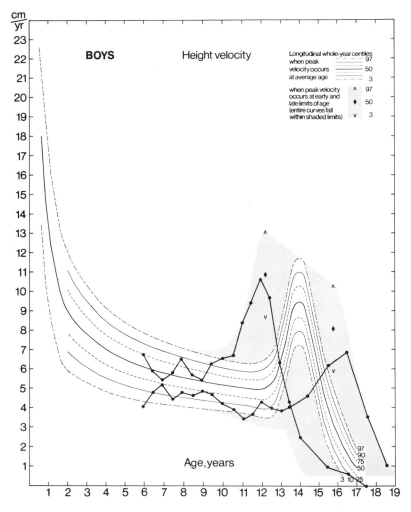

Figure 66 Whole-year height velocity plots of an early- and a late-maturing boy (From Tanner and Whitehouse, 1976)

and Davies, 1985), which are similar to the British charts except that the use of colour printing makes possible the representation, even in the distance charts, of the 50th centile lines for the 2SD-early and the 2SD-late boys and girls, together with the 95th centile (in this instance) for 2SD early and the 5th centile for 2SD late. Figures 67, 68, 69 and 70 show these charts, but unfortunately without the colour so that they look much more

HEIGHT

Figure 67 Height attained standard for North American boys. *(Red) lines:* 50th centile (solid) and 95th centile (dashed) for boys 2SD of tempo early; *(Green) lines:* 50th centile (solid) and 5th centile (dashed) for boys 2SD of tempo late
(From Tanner and Davies, 1985)

HEIGHT

Figure 68   Height attained standard for North American girls. *(Red) lines:* 50th centile (solid) and 95th centile (dashed) for girls 2SD of tempo early; *(Green) lines:* 50th centile (solid) and 5th centile (dashed) for girls 2SD of tempo late (From Tanner and Davies, 1985)

HEIGHT VELOCITY

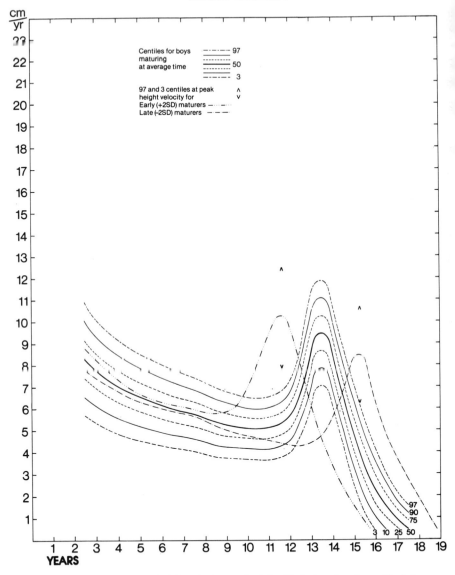

Figure 69   Whole-year height velocity standards for North American boys. *(Red) line:*
50th centile for boys 2SD of tempo early;*(Green) line* : 50th centile for boys 2SD of
tempo late. ∧ and ∨, 97th and 3rd centiles for peak velocities of early-and late-maturers,
respectively
(From Tanner and Davies, 1985)

HEIGHT VELOCITY

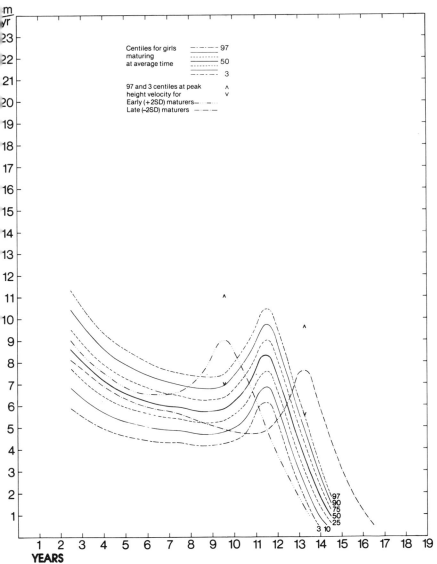

Figure 70   Whole-year height velocity standards for North American girls. *(Red) line:* 50th centile for girls 2SD of tempo early;*(Green) line* : 50th centile for girls 2SD of tempo late. Λ and V, 97th and 3rd centiles for peak velocities of early-and late-maturers, respectively
(From Tanner and Davies, 1985)

confusing than they really are. (Like British charts, they are available from Castlemead Publications, 4a Crane Mead, Ware, Herts, SG12 9PY, England. Both these and the British charts for actual use have background grids printed in, removed here for increased clarity.)

Weight Charts

British weight distance charts are given in Figures 71 and 72 and the weight velocities in Figures 73 and 74. Delaying their appearance till now

Figure 71　　Standards for weight: boys　　　　　(From Tanner and Whitehouse, 1976)

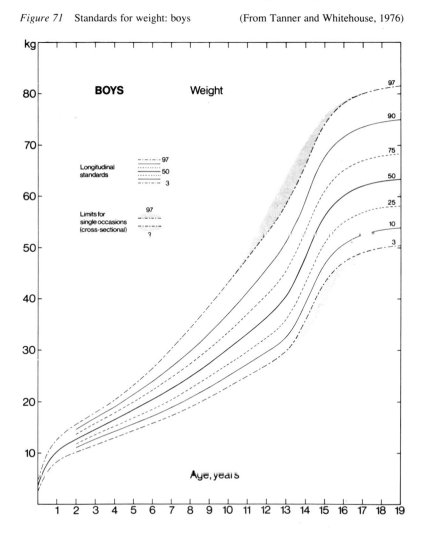

is deliberate. Because it is easy to measure, weight has come to have a quite disproportionate importance in the minds of many people. But because weight represents the sum of so many different tissues, it is harder to interpret a weight chart than a height one, and in the growth-disorder clinic weight charts are much less important. Weight may be a good measure of obesity, if, and only if, it is considered in relation to height; but skinfold thickness (*see* Chapter 1, p. 18) is a much better one.

Weight changes are important as an index in acute illness or starvation, of course, but in chronic long-term growth disorders height is more to the

Figure 72   Standards for weight: girls   (From Tanner and Whitehouse, 1976)

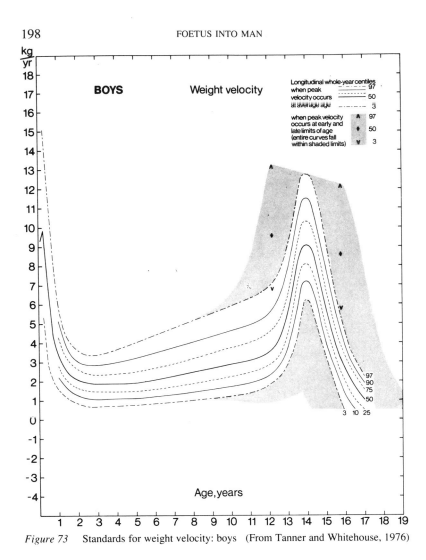

Figure 73   Standards for weight velocity: boys   (From Tanner and Whitehouse, 1976)

point. Weight can be very misleading sometimes. For example, a child with growth hormone deficiency becomes fat as well as small, and when treated with human growth hormone he loses fat and gains height. In consequence, he may actually lose weight to begin with. Similarly when treatment is stopped he gains weight, but this is no sign of continuing growth; on the contrary, growth in height ceases.

Charts of weight for given height, covering all ages mixed together, are often used by nutritionists in surveys (*see* Figure 81). They are attended

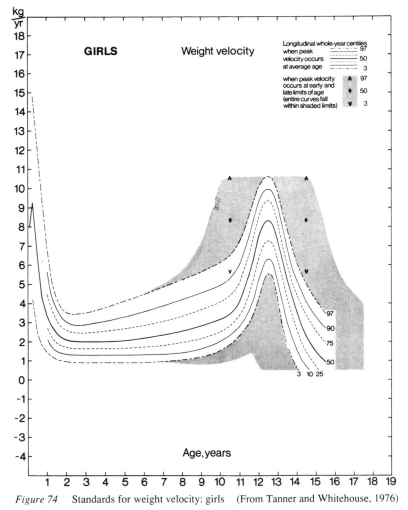

Figure 74   Standards for weight velocity: girls   (From Tanner and Whitehouse, 1976)

by the problem, however, that the centile distribution of weight for given height is not the same for children of one age as for children of another. Thus in interpreting weight for height, it matters what age the child is, and it matters more at some ages than at others. The only safe course is to have charts of weight for height for each age, and these have been given by Chinn and Morris (1980) for ages 5 to 11. At puberty the situation is worse still owing to variations in puberty stage, which also have to be allowed for.

Another way to approach this problem is to use an index such as weight/height[2]. Ratios like this have problems of their own, which take us beyond the scope of this book. The paper by Cole (1985) is a valuable introduction,

Charts from Birth to Five Years

The charts for height and weight so far presented are not altogether convenient for young children because there is little room for detailed plotting. Figures 75 and 76 are consequently recommended for use as

Figure 75 *(a)* Standards for length–height; ages 0-5; boys
(From Tanner and Whitehouse, 1973)

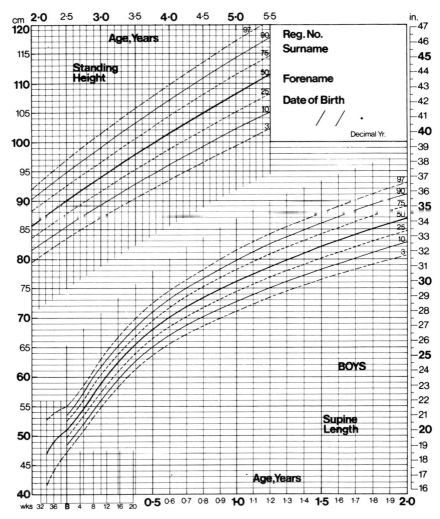

distance charts in this age group. They also have the advantage that children born pre-term can be accommodated on them. A child born at 36 weeks instead of 40 should not, 6 months after birth, be plotted at age 6 months, or 26 postnatal weeks, but at 22 postnatal weeks. Failure to do this can result in considerable error in the first year after birth. Two scales are used in these charts, one from birth to 2 years, and a less expanded scale from 2 to 5 years (the top left sections). The standardizing data are the same as in the 0–19 charts.

Figure 75   *(b)* Standards for weight; ages 0–5 boys
(From Tanner and Whitehouse, 1973)

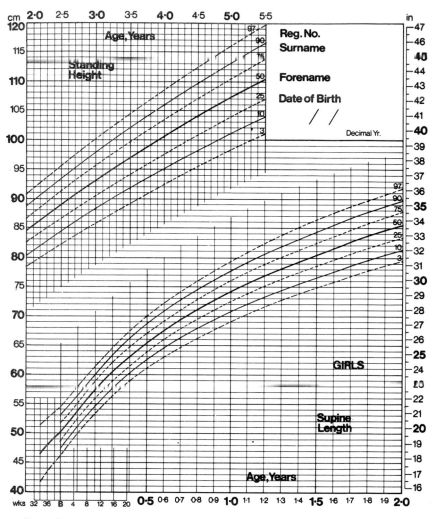

Figure 76 (a) Standards for length–height; ages 0–5 girls
(From Tanner and Whitehouse, 1973)

United States Population Charts

Until recently only cross-sectional charts were available for North American
children. These are issued by the National Center for Health Statistics
(NCHS). They are based from birth to 3·0 years on the relatively limited
and somewhat selective data of the Fels Research Institute collected in the
1960–75 period. From 3·0 to 18·0 years, however, the data come from two

Figure 76 (b) Standards for weight; ages 0–5 girls
(From Tanner and Whitehouse, 1973)

specially mounted surveys (Health Examination Survey, 1962–70, and Health and Nutrition Examination Survey, 1971–74). These surveys aimed to sample the whole child population of the USA in a proportionate way, i.e. with numbers in each geographical area proportionate to the child population there. The resulting standards for height and weight are given in Figures 77 to 80, reproduced from the paper by Waterlow *et al.*, 1978.

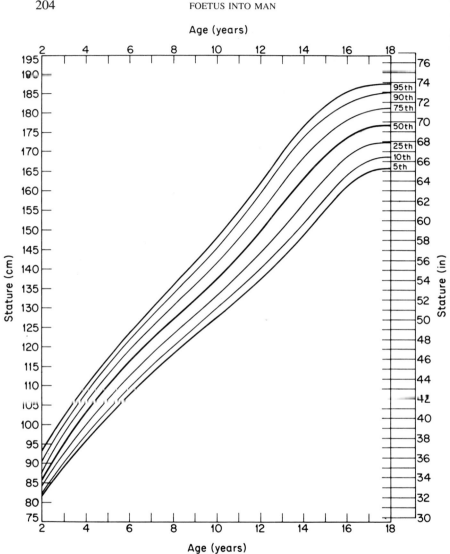

Figure 77   Population standards for height for USA boys
(Data from National Center for Health Statistics. Cross-sectional-type charts)

The 5th and 95th centiles are given rather than the 3rd and 97th, because
the numbers varied from 300 to 600 children in each yearly age group and
the errors of locating the outside centiles are large unless the sample is a
very big one.

Figure 78 Population standards for height for USA girls
(Data from National Center for Health Statistics. Cross-sectional-type charts)

It must be remembered that these are *distance* standards founded on cross-sectional data reported simply, and not used, to give limits around the true longitudinal-shaped curve. Thus they are appropriate for comparing populations (for undernutrition, etc.) rather than for plotting

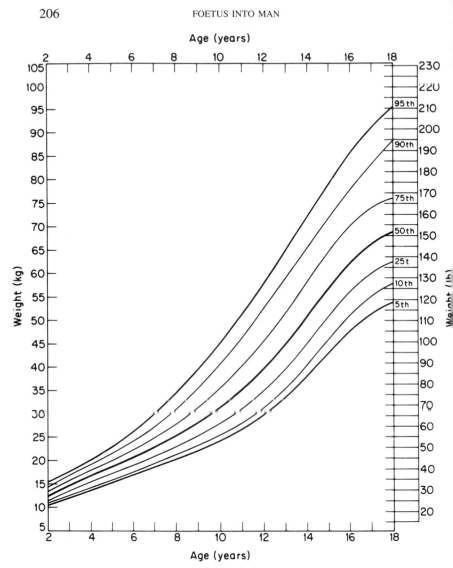

Figure 79   Population standards for weight for USA boys
(Data from National Center for Health Statistics. Cross-sectional-type charts)

individuals. Indeed they were designed specifically for the former purpose, and have been recommended for use as a WHO reference for use in monitoring the nutrition of countries or subpopulations within countries.

Figures 81 and 82 show weight-for-height standards derived from the

Figure 80   Population standards for weight for USA girls
(Data from National Center for Health Statistics. Cross-sectional-type charts)

same source. These standards only apply to prepubescent children (of whatever age) and are subject to the criticism that even before puberty weight for height does depend somewhat on age. Thus the charts of Chinn and Morris, referred to above, are really preferable; but the age related

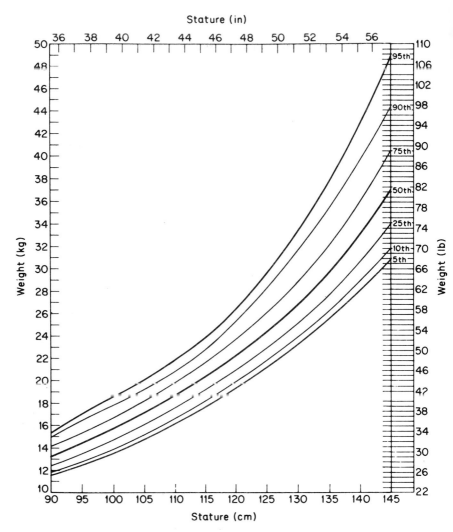

Figure 81   Standards of weight for height, irrespective of age, for pre-pubescent boys,
over the age of 1 year
(Data from National Center for Health Statistics)

error is not large between about 2 and 9 years. It has to be remembered,
also, that the North American children are fatter than most others, and that
an optimal weight for height (optimal in the medical sense of healthiest)
is probably well below the 50th centile.

Figure 82 Standards of weight for height, irrespective of age, for prepubescent girls, over the age of 1 year
(Data from National Center for Health Statistics)

Local versus Universal Standards

The WHO-sponsored use of a universal, at it happens American-derived, reference for identifying various populations around the world in terms of '30% below 5th centile for height, 20% for weight' has raised numerous critical eyebrows. It would, for example, lead to the absurd result of calling stunted through undernourishment a huge number of very healthy Japanese. Similarly the Nilotic tribesmen of the Southern Sudan, who are tall and lean, would be accounted well nourished throughout part childhood (because tall) but in the midst of present acute undernutrition (because of their low weight for height). Clearly, where money and organization permit, it is better to have local standards, which have to be updated every so often, even in the industrial world. In Czechoslovakia, for example, there were surveys in 1951, 1961, 1971, and 1981 (Prokopec 1985); in Holland in 1955, 1965 and 1980 (Roede et al., 1985).

Particularly is this so when the standards are being used as an individual screening device to pick up individuals for treatment, rather than a yardstick for looking at different subpopulations. Van Loon and his colleagues in Belgium have underlined the difficulties of using a non-local reference for this purpose and Johnston and his colleagues (1984) have pointed out the same problem – that the relationship between two measurements, in this case skinfolds, in the universal reference population and the local one may not be the same, either for genetic, or indeed ecological reasons. Even in the poorest countries, it is often possible to generate local standards, as Brown and his colleagues showed particularly well in Bangladesh.

When this is done, however, what group should be used as a basis for the standard? There is a long-running discussion as to whether one should try to assemble data from only the best-off group of the population (an 'elite' standard) or a more randomly chosen group free from overt disease (as in Van Loon et al.,). In industrialized countries this is not a serious concern; nobody suggests scrapping the current British standards in favour of ones derived exclusively from Classes I and II. But in the Third World the situation is somewhat different. It is only amongst the upper social classes that one can find, perhaps, children who are reasonably free from undernutrition, infection and parasites. In this context, use of an elite standard seems reasonable. Yet there are circumstances where it does more harm than good. So long as it is used as the standard to which other groups in the (supposed ethnically homogeneous) population should aspire and be brought to all is well. But if it is used as a standard for screening individuals, then one runs the risk of ascertaining enormous numbers of the population as sick or undernourished, and because of this neglecting the relatively few who are most in need of clinical help. Random-sample standards derived from not-overtly-ill children may serve this purpose better, provided they are continually updated.

Parents-allowed-for Charts

Figure 83 represents a different sort of chart. We have already seen how children resemble their parents, and clearly a small child with small parents represents a quite different situation from that of an equally small child with large parents. We can use the chart to allow for the heights of the parents and thus reach a more accurate assessment of whether the child has grown normally or not.

The chart was constructed with the help of the writer's colleagues R. H. Whitehouse and H. Goldstein. The data used were for parents and children from the whole range of the International Children's Centre Longitudinal Growth Studies, a group of linked studies in Brussels, London, Paris, Stockholm and Zurich, in each of which a cohort of children was followed from birth to maturity using the same techniques both for physical and psychological assessment (*see* Tanner 1981). The chart is valid only for children aged 2·0 to 9·0 years; before 2, the relationship to parental height is not sufficiently developed, and after 9·0 the adolescent spurt makes things considerably more complicated.

Statistical analysis reveals that each parent on average is equally important in determining the child's height. Despite all popular belief to the contrary, there is no closer relation of father with son or mother with daughter than father with daughter and mother with son (*see* Chapter 9, p.124). Consequently, we may use the simple average of parents' heights (uncorrected for sex) with which to compare the height of the child. This average is called mid-parent height and is placed on the horizontal scale of Figure 83 with indications of the 3rd, 50th and 97th centile of mid-parent height in the UK population.

The form in which a child's height is related to mid-parent height can be expressed as what is called a regression equation. Thus the height H of a boy may be estimated as: $H = a + bX$ where X is mid-parent height and a and b are constants. Statistical technique enables us to calculate the likely range of variation around this estimate in a way exactly similar to the centiles we have already met. A girl with parents of heights 160 cm and 180 cm has a height given by $H = a + 170b$, where a and b depend on the girl's exact age. If the girl is 8·0 years H is estimated as 127 cm. The 3rd and 97th centile limits to this estimate are 118 cm and 136 cm. We now see how tall the girl actually is. If she is 116 cm she is below the 3rd centile for her parents' heights even though this value does not carry her below the 3rd centile for the population at large.

The chart is constructed so that the child's height centile has first to be looked up in the regular charts of Figures 60 and 61. In this way Figure 83

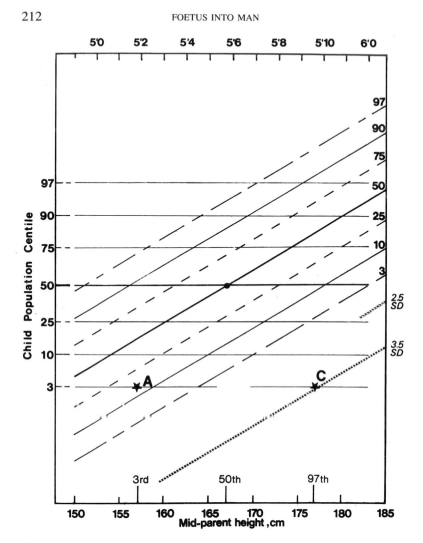

Figure 83   Standards for height at ages 2·0 to 9·0, allowing for height of parents. For child population centile *see* Figures 60 or 61; centiles may be interpolated by eye. A and C are two children both at 3rd centile for population, with mid-parent centiles at 3rd and 97th

may be used for both boys and girls. The regular centiles constitute the vertical axis of the parent-allowed-for graphs; the exact centile can be estimated by interpolation with sufficient accuracy for our purpose of assessment. Suppose we have a boy at the 3rd centile in Figure 60. His parents are very small, with mid-parent height itself at the 3rd centile (157 cm). Draw an imaginary line across the graph from the 3rd centile point on the vertical axis and another up the graph from the 3rd centile point on the horizontal axis; plot the point where the lines cross (i.e. A in the figure). Now consider A in relation to the system of diagonal lines, which represent the regression equation standard. The lines are marked in centiles (with indications of two lower ones at 2·5 and 3·5 standard deviations which may be ignored by the statistically innocent). The point A lies between the 10th and 25th lines, at about the 15th centile. The interpretation is straightforward. Though the boy is at the 3rd centile for the population of all boys of his age, he is at the 15th centile when his parents' heights are taken into account, and thus well within the limits of normal for his family. (Clearly, this interpretation, and indeed the whole use of this approach, is only valid if the parents have grown to their full height potential or somewhere near it. If the parents' heights have been curtailed by a poor environment the correlation between parents and children breaks down, as explained in Chapter 9. The chart also does presume that the advertised parents are, indeed, the real ones.)

Consider a second boy, also at the 3rd centile for his population but with very large parents whose mid-parent centile is at the 97th (177 cm). The resulting point is labelled C in the figure. Here the situation is radically different; this boy is far below the 3rd centile for his family and in the clearly pathological range. He should be investigated (the child studied was one with asymptomatic coeliac disease). Rather than plot the intermediate point between A and C representing the 3rd centile boy with parents at the mid-parent 50th centile, let us leave readers who believe their statistical innocence has been lost, the question: 'Why is the intermediate point below, not on, the 3rd regression centile?'

Puberty Standards

Besides these standards for growth and growth velocity we can construct standards for assessing whether the various stages of puberty have been attained at the usual ages. The stages of breasts in girls, genitalia in boys and pubic hair in both sexes are illustrated in Figures 84 to 87. Extended descriptions need not be given here; suffice to say that breast stage 1 (B1) represents pre-adolescence, B2 the first appearance of what is called the breast bud, B3 a small pubescent breast, B4 a stage where the areola projects beyond the contour of the breast (a stage skipped by some girls and persisted in for several years by a few), and stage B5 the mature pre-pregnant breast. Stages of pubic hair are the same for both sexes, with

Figure 84   Breast standards   (From Tanner, 1962)

extension of hair up the abdominal mid-line in the male not counted in the system.

The terminology is important; B2 signifies the very first moment when B2 becomes visible, technically the transition from B1 to B2. The stage B2 + lasts from B2 to B3. The girls' puberty-stages chart (Figure 87 lower) can now be readily understood. We cannot provide centiles for breast development at each age like those for height because we cannot measure the breast; we can only rate its appearance in terms of separate stages.

Figure 85 Genitalia standards   (From Tanner, 1962)

Therefore we turn the standards round 90°, so to speak, and instead of asking 'Given this girl's age, is her height within normal limits?' we ask 'Given this girl's breast appearance, is her *age* within normal limits?' The standards tell us that the average (50th centile) girl with the breast appearance B2 is aged 11·2 years; the early limit for such an appearance (quite arbitrarily labelled 97th centile rather than 3rd) is 9·2 years and the late limit is 13·2 years. Subsequent stages are plotted in the same way except that no age limits are possible for stage 5 since it persists indefinitely.

Figure 86   Pubic hair standards: (*a*) boys; (*b*) girls   (From Tanner, 1962)

The example shows ratings for a girl followed longitudinally. Her breasts developed considerably later than average, and her pubic hair slightly later; her menarche occurred at almost the late limit of the normal range.

The boys' charts (Figure 87, upper) is similar, except that the two lines at the bottom represent the attainment of two particular sizes of the testes as measured by palpation and comparison with the models of the Prader orchidometer (*see* Chapter 5, p.58). Out of the progression of sizes these two are chosen because 4 ml represents the first definitive enlargement at

Figure 87 Standards for puberty: (*a*) boys; (*b*) girls. One individual of each sex, serially followed through puberty, is plotted
(From Tanner, 1962)

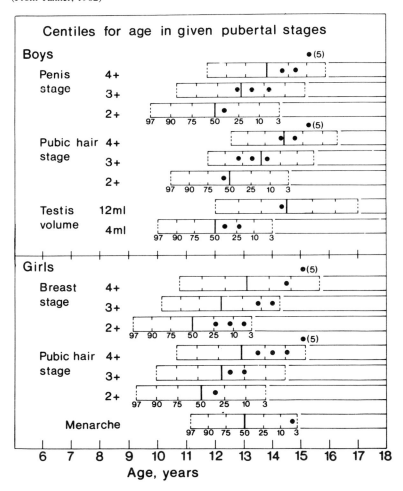

puberty and 12 ml is a size reached at maturity by everyone (though some attain 20 or 25 ml).

Standards for Bone Age

It is beyond the scope of this book to give standards for skeletal maturity or bone age. The left wrist and hand are X-rayed, and the appearance of the bones compared with standard appearances given in the Tanner–Whitehouse or Greulich–Pyle systems (see Chapter 6, p. 76). What emerges in the Tanner–Whitehouse system is a score for maturity. Maturity differs in an important way from a measurement such as stature, in that the normal growth process takes every individual from a common condition of being wholly immature to another of being wholly mature. Maturity can thus be measured on a scale of, say, 0 to 100, for everyone. An individual's score can then be compared with the centile distribution of scores of other individuals of the same age in just the same manner as height. We can then say whether skeletal maturity is advanced or delayed.

Alternatively, we can convert this maturity assessment into a bone age: the age at which the average child of the standard population has this particular maturity score. Thus if a boy has a maturity score of 60 units at age 12·0, one looks up a table and discovers that 60 is the score shown by the *average* boy at 10·0 years. The boy in question is therefore said to have a bone age of 10·0 'years', i.e. one 2 years retarded in relation to his chronological age. At ages greater than 3·0 years the limits of normal bone-age delay and advance are about 2 years in either direction. Beyond this, pathology may be expected. The limits are narrower from 2 to 3 years and before age 2 skeletal maturity derived from the hand and wrist is inaccurate owing to lack of sufficient bones in the X-ray; radiographs of knee and ankle have to be used instead.

Prediction of Adult Height

A child's adult height may be predicted from the height of its parents, the most likely height centile being the average of the height centiles of mother and father (or, more precisely, the average of standard deviation-scores of mother and father). Such a prediction can be made even before the child is born. The error is fairly great (represented by a correlation coefficient of only about 0·7) and a better prediction can be made from the child's own height, at least after the age of 2. From age 2 until the beginning of puberty the correlation between the child's present height and his adult height is between 0·80 and 0·85, which means that 95% of the time the error of prediction will be within the limits of about 15 cm (6 ins) either side of the predicted value. Though this may not sound very accurate, it does represent a considerable restriction compared with the whole population range which is ± 25 cm about the average for men and ± 22 cm for women. It is better, too, than prediction from parents' centiles, which gives an error of about ± 18 cm.

The simplest way to set forth these predictions is by giving a table of the average percentage of adult height reached at each age. Table 5 gives values for both the British and North American standard longitudinal data. The table shows very well the advancement of girls as compared with boys and vindicates the old wives' tale that a boy doubles the length he has on his second birthday, but girls more nearly double the length they have at 18 months. The prediction represents the average for all children for those ages placed in the central block, i.e. 2·0 to 10·0 years for boys and 2·0 to 8·0 years for girls. Beyond 10 in boys and 8 in girls the percentages apply only to those boys and girls who have their adolescent growth spurts at exactly the average time.

During adolescence these simple predictions become much less accurate because they fail to take into account whether the girl is an early maturer and hence at a given age has little more growth to occur, or a late maturer with the full adolescent spurt yet to come. During this period it is essential to include in the prediction a measure of bone age. When this is done the prediction becomes quite accurate again. Then the limits within which 95% of predictions fall are, for boys, ± 7cm up to age 12, ± 6 cm at ages 13 and 14 and ± 5 cm at age 15 and 16. For girls, the errors are approximately ± 6 cm up to age 11, ± 5cm at age 12 if the girl has not yet had menarche but only ± 2 cm is she has passed menarche, ± 4 cm at 13 if premenarcheal and ± 2cm if postmenarcheal, and ± 2 cm at age 14.

When bone age is used the prediction is made through regression equations specific for each age. As an example, consider a boy aged 11·0 years. His bone age on the Tanner–Whitehouse system ('RUS bone age', meaning bone age based on radius, ulna, metacarpals and phalanges only) is, say, 12·2 'years'. The equation for predicting his adult height is:

Adult height = 1·11 (present height)
$\qquad$ -1·85 (chronological age) -7·8 (bone age) + 3·6

Tables of the regressions and full details of their use will be found in *Assessment of Skeletal Maturity and Prediction of Adult Height* (Tanner *et al.*, 1983). Prediction may also be made by use of the Greulich–Pyle

TABLE 5   (a) Percentage of adult height reached at successive ages by average boy and girl in UK. *Values during puberty (below lower horizontal lines) refer only to children with peak height velocity occurring at average time (12·0 and 14·0 years) (b) Percentage of adult height reached at successive years of age by boys and girls in the USA*

Age (yr)	[a] UK Boys	[b] USA Boys	[a] UK Girls	[b] USA Girls
0·08	30·9	30·8	32·7	32·6
0·25	34·7		36·4	
0·50	39·0	38·5	40·4	40·3
0·75	41·6		43·3	
1·00	43·7	43·0	45·7	45·3
1·25	45·4		47·8	
1·50	47·0	46·7	49·7	49·3
1·75	48·4		51·3	
2·00	49·8	49·4	52·8	52·7
2·00	49·2	49·2	52·1	52·5
2·5	51·6		54·8	
3·0	53·9	53·9	57·3	57·8
3·5	56·1		59·7	
4·0	58·2	58·1	61·9	62·4
4·5	60·1		64·0	
5·0	62·0	61·9	66·1	66·5
5·5	63·8		68·0	
6·0	65·6	65·6	69·9	70·5
6·5	67·3		71·8	
7·0	69·0	68·9	73·6	74·2
7·5	70·6		75·3	
8·0	72·2	72·2	77·1	77·0
8·5	73·8		78·8	
9·0	75·4	75·2	80·5	81·3
9·5	76·9		82·2	
10·0	78·3	78·2	83·8	84·8
10·5	79·8		85·5	
11·0	81·3	81·1	87·3	88·9
11·5	82·7		89·5	
12·0	84·1	84·1	92·2	93·6
12·5	85·5		94·7	
13·0	87·1	87·9	96·7	97·6
13·5	89·1		98·0	
14·0	92·0	93·0	98·9	99·4
14·5	94·6		99·5	
15·0	96·6	96·7	99·8	99·9
15·5	97·9		99·9	
16·0	98·8	98·5	100·0	100·0
16·5	99·4		—	
17·0	99·8	99·4	—	
17·5	99·9		—	
18·0	100·0	100·0	—	

(a) Based on Tanner–Whitehouse longitudinal standards
(b) Based on cross-sectional standards of NCHS

atlas of bone age, combined with tables provided by Dr Nancy Bayley, a psychologist by profession, whose papers on physical growth in the period 1930–60 rank with those of the two American 'immortals' Franz Boas (anthropologist, papers 1880–1930) and Frank Shuttleworth (another psychologist, papers 1930–50). The more modern method is, however, somewhat the more accurate.

Prediction of adult height is not just an amusing pastime. In certain jobs height is a crucial qualification, and for many years the Department of Growth and Development at the Institute of Child Health advised the Royal Ballet School as to the predicted height of children seeking to enter its school, usually at age 9 to 10. Ultimate entry to the corps de ballet is restricted to girls and boys within fairly narrow ranges of heights, so if the prediction clearly indicates that an entrant's chance of ending up within these limits is slender then both the child and the parents are told, and advised that though teaching, or other styles of dancing, may be possible, classical ballet is not. In this way, a terrible disappointment at the end of years of specialized training may be avoided.

Prediction of adult height is often most reassuring to tall girls believing, usually wrongly, that they are going to grow into 'giantesses', and to short boys, usually with considerable growth delay, who fear the opposite. The predictions do not apply, however, to children with endocrine or bone pathologies affecting growth.

CHAPTER 12

Disorders of Growth

This final chapter gives brief descriptions of the common disorders of growth, in terms shorn, so far as possible, of medical terminology. Growth disorders are at present claiming an increasing share of the family doctor's and paediatrician's attention as the infectious diseases of childhood are brought under control, and in the UK and USA specialised growth-disorder clinics have been set up in numerous university medical centres, equipped with proper instruments for measuring children and assessing their maturity, and staffed by persons trained in auxological methods. Such clinics have close working arrangements with departments of paediatrics, endocrinology, medical genetics, orthopaedics and psychiatry. There are several associations of parents of short-statured children. In the UK the Child Growth Foundation[1], and in the USA the Human Growth Foundation[2], are concerned principally with endocrine causes of short stature and the various categories treatable with growth hormone. Disorders of bone and its various syndromes (*see below*) are the particular concern of the UK's Association for Research into Restricted Growth (ARRG) and the Little People of America (LPA).

Causes of Short Stature

Most of the children attending growth clinics have short stature as their main complaint (we never use the word 'dwarfism': to many parents and children it has a curiously magical ring, derived from an amalgam of medieval folk-tales, gargoyle grotesques, Wagnerian Nibelungen and Scandinavian troll-kings. It is a term better totally avoided).
Short stature may be due to:
1 Normal genetic shortness
2 Growth delay
3 Chromosomal disorders
4 Intra-uterine maldevelopment

[1] UK
  Child Growth Foundation, 2 Mayfield Avenue, Chiswick, London W4 1PW
  Association for Research into Restricted Growth, 61 Lady Walk, Maple Cross, Rickmansworth, Herts WD3 2YZ
[2] USA
  Human Growth Foundation, The Montgomery Building, 1720 Montgomery Lane, Bethesda, Maryland 20814
  Little People of America, Inc, PO Box 633, San Bruno, CA

222

5   Endocrine gland disorders
6   Cartilage and bone disorders
7   Disorders of absorption of food
8   General diseases of the lungs, heart and kidney
9   Steroids and radiation
10   Psychological causes

These will be discussed in turn, followed by tall stature.

1. Normal Genetic Shortness

Standards for children's height based on the whole population of children are only a first convenience. If a child is short, i.e. by convention below the 3rd centile, he should be assessed (if between 2·0 and 9·0 years) on the more accurate standards which allow for his parents' heights, as described in Chapter 11. If he is within normal limits for his parents, then he himself is normal, whatever the population position unless, of course, one or other of his parents has a growth disorder or has been brought up under economic conditions restricting growth. Note that both parents should be seen and actually measured. Estimates of height are notoriously inaccurate, particularly estimates by wives of their husbands' heights (they usually increase it: sometimes the reverse).

Families of short people probably (though the matter is currently in dispute) secrete no less growth hormone than families of tall ones. More likely the difference resides in the reaction of the growth cartilage cells either to growth hormone, or, perhaps, to their own self-stimulating product somatomedin or IGF -1 (*see* Chapter 7 p. 89). In either case however it is possible that increasing the amount of growth hormone circulating in the blood to supranormal levels might induce extra activity of the cartilage, and hence extra growth. Current clinical trials seem to indicate that extra, injected, growth hormone in many though not all such children causes growth to be accelerated at least for one or two years. Beyond this we have no data. Nor do we have any really trustworthy data on the most crucial point – whether such treatment, extended over all the years of growth, would actually result in an increased final height. If it did so, that might by some be construed a satisfactory and socially justifiable 'treatment'.

But the biologist must point out that what the 'treatment' would have produced is biological pathology, a state defying the programming of the genes, a sort of mirror image of chronic malnutrition. And if all 3rd centile children (all, that is, whose parents could afford the currently very high cost) can be turned into 25th centile ones, what will those at the 10th centile say?

2. *Growth Delay*

This is a very common cause of attendance at a growth disorder clinic: thus every child attending has his bone age assessed. Delay may be manifest in early childhood as well as at the age when normal puberty should occur. Delay of up to 2 years is normal in the sense that it occurs in 3 out of every 100 healthy children; smaller numbers of children are delayed 3 or even 4 years. Frequently one or other of the parents, or an aunt or uncle or brother or sister, has also been delayed. The mother is asked about her age at menarche, which may well be 15 or 16 years instead of the more usual 12 to 14 years. A history of delay in the father's puberty can sometimes be obtained (on occasion from his wife) but in general this is more difficult owing to the lack of a prominent landmark like menarche.

Growth delay in itself is not abnormal: puberty occurs in the end, and develops in every way quite normally. Thus boys and girls with this diagnosis can be absolutely reassured, with a full demonstration of figures and facts, charts and predictions. Children find the information that they are simply taking after their mother or father reassuring, if not particularly welcome. What reassurance does not do is remove the very real physical handicap associated with being small and even more with being prepubertal in a society of adolescents. The situation may be compounded or precipitated by a younger brother or sister who ploughs on ahead while the delayed child hangs around at the lowest growth velocity he or she has ever experienced (the immediately pre-adolescent velocity being so low that parents sometimes allege, mistakenly, that the child has actually stopped growing altogether for the last months or year).

Parents, child and doctor may be tempted to give hormonal treatments to accelerate growth and to bring about the outward signs of puberty, particularly in boys. Testosterone is indeed effective for this: the problem is that if the dose is too great it accelerates bone age to a greater relative degree than height. If this occurs the final height of the boy is diminished, and this is too high a price to pay for the immediate acceleration, particularly as growth-delayed children who come to the clinic tend also to be genetically somewhat small, presumably because these are the ones who feel the pinch most. However, if the dose is kept low, bone age is only accelerated to the same degree as height and then there is no diminution in final height. Testosterone-like synthetics (anabolic steroids) may also be used: oxandrolone is the best researched of these but is not passed for prescription at present in the UK. Low doses of these steroids produce some increase of growth velocity without bringing about the growth of penis and pubic hair that testosterone induces, so that they can be used when growth delay is recognised some years before the pubertal age period, and all that is required is an increase in growth tempo for a few years.

The treatment of girls with growth delay is more contentious, and at present much less sought after. Low doses of oestrogen will often accelerate growth, and usually bring about breast development also. If any of these hormonal treatments are undertaken they should be carefully controlled by a physician with specialist training in auxological problems. Figure 88 shows a typical growth curve of a child with growth delay, compared with curves of children with other diagnoses.

Some paediatric endocrinologists think that growth delay is due to a sort of micro-deficiency of growth hormone; the delayed child is supposed to have secreted, over his whole childhood, slightly smaller amounts of growth hormone than average. If so, administration of growth hormone would be a logical treatment. Other paediatric endocrinologists do not agree. It does seem that growth-delayed children (like short, not-delayed ones) can have their growth velocity increased by giving them extra amounts of growth hormone but so far we only know that this works in the short term, over a year or two. Whether such treatment, if continued over five or ten years, would produce an increase in adult height is as yet unknown.

3. Chromosome Disorders

A number of chromosome disorders produce short stature, often in conjunction with other anomalies. A well-known instance is Down syndrome (also called mongolism because of the appearance of the eyelid) in which the child has 3 instead of 2 chromosomes N° 21. In this instance the child is mentally handicapped as well as small. The appearance and behaviour makes diagnosis at an early age relatively easy: it is confirmed by examination of the karyotype (*see* Chapter 4, p. 53)

Another relatively common chromosome disorder, much milder in its effects, is Turner syndrome, confined to girls, and due to a lack of one of their normal pair of chromosomes. Girls with Turner syndrome are not mentally handicapped; in addition to their short stature they may or may not have special physical features such as puffy hands and feet soon after

Figure 88   Typical height curves of girls with growth delay (small/delay), Silver–Russell
syndrome, Turner syndrome and isolated growth-hormone deficiency
(From Tanner, Lejarraga and Cameron, 1975)

birth, webbing of the neck, broad chest, small nipples and a somewhat
characteristic face. The bone age is normal or only slightly delayed until
the time puberty should occur. But then, in the full blown case, the breasts
fail to develop and the adolescent growth spurt does not occur because in

these patients the ovaries fail to develop and no oestrogen is secreted. Sometimes pubic hair develops sparsely, sometimes not at all. The diagnosis is confirmed by a karyotype. No treatment can make good the lack of ovaries and the news of life-long infertility has to be broken to parents and child. The breasts can be made to grow by giving oestrogen, which sometimes is accompanied by pubic hair growth as well. An artificial 'menstruation' can be produced by giving oestrogen for 3 weeks followed by a week without treatment or with a progesterone-like hormone. These girls commonly wish to menstruate like their friends, at least until the time when their friends begin to give it up. In the last few years it has been shown that growth hormone, alone or in combination with oxandrolone, increases growth velocity substantially in Turner syndrome, and present indications are that at least in the average patient actual final height is increased too. In the absence of growth hormone treatment, final height averages about 145cm (4ft 9in), though it depends on the height of the parents, since the parental correlation seems, curiously, to be fully maintained.

The above describes the situation in a full Turner syndrome, karyotype written XO. Mosaics (*see* p. 52) also occur in which only a proportion of cells have this karyotype, others being the normal female XX. In such cases the symptoms are reduced and all gradations from normal to the complete Turner syndrome are seen, both as regards sex development and stature. It follows that any girl with short stature of unknown cause should have a full karyotype done, with a substantial number of cells examined for their chromosome complement. Sometimes diagnosis may actually be impossible, since a minor XO cell line may produce some small effects and then die out, leaving only XX cells.

4. Intra-uterine Maldevelopment

Maldevelopment in utero may be due to a fault in the fertilized ovum, a fault in the placenta limiting the supply of foods and oxygen to the foetus, or to disease or starvation of the mother. Certain of the first and all of the second and third sorts of maldevelopment cause a reduction of growth and small size at birth. This smallness of size may be known only through weight. Though length would often be as or more informative, it is unfortunately not routinely taken in clinics in the UK although accurate apparatus for measuring it now exists.

Standards for birthweight are available (*see* Bibliography); the centiles are different for girls and boys and for first-born and later-born – and, of

course, for each week of gestational age. We are concerned here only with
babies who are below the centile limits for weight at their particular
gestational ages. Babies who are simply born early at the normal weight
for that length of gestation seem mostly to catch up to perfectly normal
height. Small-for-gestational-age babies may be born at term, or may be
born after a shorter gestational period.

The majority of such small-for-gestational-age children, particularly
those born at term, develop within the normal centiles, though they do not
fulfil their genetic potential, and hence appear somewhat small for their
parents' heights. Their average centile seems to be about the 30th for the
population. However, some fail to catch up in this way. These children
remain short and lacking in subcutaneous fat and have a characteristic
facial appearance. The face as a whole is triangular, with large eyes and
a small lower jaw; the forehead is large and prominent in relation to the
face, and the ears are set low in the head and tend to protrude. The bridge
of the nose is usually depressed and the mouth turned down at the corners
(giving rise to the expression 'sharkmouthed'). A proportion of these
children have marked asymmetry of the limbs or body, one arm or leg
being longer than the other, or one side of the face or chest more
developed. This condition constitutes a specific syndrome, known by the
name Silver–Russell after the two paediatricians who independently
described it in the 1950s. Mental development is usually normal.

These children grow up normally in every way except for size. They are
healthy and active and should not be regarded as delicate. In middle
childhood they put on some subcutaneous fat and their puberty occurs at
the normal time and with the normal sequence. The syndrome hardly ever
occurs twice in the same family; thus the mother of such a child can be
reassured about subsequent pregnancies. What the cause of the disorder
is, we do not know. The prognosis may be affected by the treatment given
to low birthweight babies nowadays, which consists of feeding them more
intensively than before, and keeping them warmer. Extra growth hormone
makes many of these children, like those with delay, grow faster and may
– though this is still uncertain – increase final height.

Some of the children with relatively low birthweights and subsequent
short stature have rarer and more specific syndromes of maldevelopment
(such as de Lange syndrome, Prader–Willi syndrome and the syndrome
due to excessive intake of alcohol by the mother during pregnancy. There
are very many such syndromes with only a few children suffering from
each. Most of the syndromes are probably due to intrinsic faults in the
fertilised ovum, and some are inherited. The subject is a highly specialised

one, only adequately tackled by paediatricians and medical geneticists with a particular interest in it.

5.   Endocrine Gland Disorders

The two chief endocrine disorders which cause short stature are growth hormone insufficiency and thyroid insufficiency. A third cause is disease of the adrenal or other glands, leading to precocious puberty: in this case the child grows fast and develops the changes of puberty pathologically early. Thus early on the child is very large for his or her age but as growth stops early the final result is short stature. Most children with precocious puberty can now be very effectively treated with a slightly altered form of gonadotrophin-releasing hormone (*see* p. 89) which actually turns off the secretion of the usual hormone, thus stopping puberty.

Growth hormone insufficiency   A chart of the growth of two brothers with insufficiency of growth hormone secretion has already been shown (Figure 57, p. 168). The insufficiency does not affect intra-uterine growth, at least to an extent measurable by birthweight, but from birth onwards growth is less than normal. This is seen in Figure 89 which illustrates the height curves of two monozygotic twins, one with isolated growth-hormone lack. Their father was an engineer and their mother a teacher of physical education, and the children were measured from birth onwards simply because of the interest of their being twins. The parents soon realized that something was wrong and eventually growth-hormone insufficiency was recognised – much later than it should have been, for the whole subject was very new at the time. If all children had their length measured in pre-school clinics most cases of this disorder could be recognised by age 2 and proper and effective treatment promptly instituted.

Most cases of growth-hormone insufficiency are called 'idiopathic' meaning 'of unknown origin'. Others are associated with tumours of the central nervous system which destroy the part of the brain controlling the pituitary gland (*see* p. 87), the pituitary itself, or both. These tumours are nearly always benign, i.e. not cancerous, and the one which is by far the most common, craniopharyngioma, is not technically a brain tumour at all but part of a piece of mouth epithelium pinched off and left behind during development. Craniopharyngiomas can be recognized in a skull X-ray or by computer-assisted tomography and most are readily removeable. Occasionally thereafter they recur, but are removeable again without great danger. The child with such a tumour occasionally comes

Figure 89 Growth of two monozygotic twins, one with isolated growth-hormone deficiency. Measurements after break in lines made in Growth Disorder Clinic. Human growth-hormone treatment, begun at arrow, has continued

to the physician with the sole complaint of short stature, but more usually the first symptoms are neurological and characteristic of a tumour in the head. After removal of the tumour the child may suffer from any combination of pituitary hormone insufficiences. Most, though not all, have growth-hormone deficiency. These children are treated with human growth hormone, often in conjunction with thyroxine, cortisol (hydrocortisone), posterior pituitary hormone and, at the time of puberty, gonadotrophins or sex hormones. A second form of growth hormone insufficiency is associated with irradiation of the head with X-rays for the treatment of some forms of childhood leukemia (where the cancerous cells are found in the membranes covering the brain).

Idiopathic growth-hormone insufficiency in its severe form occurs, it seems, in about 1 in 5,000 births in the UK population at present. This may, however, change, for one of the chief clues to its cause is its association with breech birth, and also with the use of forceps in mothers who have already had one baby (and therefore would not normally require forceps). Whether the breech birth contributes to its causation, or whether for some obscure reason, lack of growth hormone is a factor contributing to the breech position, is unknown. But as an increasing number of obstetricians now deliver by caesarian section babies who would otherwise be breech births the prevalence of growth hormone insufficiency should diminish.

There is evidently a hereditary predisposition, at least in some cases. About 3% of cases have brothers or sisters with the disorder, like the brothers affected in Figure 57 (p. 168). In a very few families one of the parents is affected and the actual gene for growth hormone is abnormal. Curiously, the deficiency is more than twice as common in boys as in girls, for reasons quite unknown.

Children with severe growth hormone insufficiency are simply small with normal skeletal proportions, facial appearance and intelligence. They are usually fat (the condition diminishes on treatment) and they have delayed bone age. The diagnosis is confirmed by lack of growth-hormone production in response to a stimulation test, such as clonidine or a sudden drop in blood sugar caused by an intravenous injection of insulin. The lack may be of growth hormone only, or other pituitary hormones may also be affected. In the majority of cases in the UK growth hormone alone is involved: the next most common associated insufficiences are those of TSH, the thyroid-stimulating hormone, and FSH and LH, the gonadotrophins. ACTH, the adrenal-stimulating hormone is much less frequently involved though some patients, who initially suffer only from growth hormone lack, gradually develop the other insufficiencies as the years go by, ending up with ACTH insufficiency in their late twenties or thirties.

Children with less severe forms of growth hormone insufficiency may respond normally to a pharmacological stimulus, but fail to produce a normal amount of growth hormone in their ordinary life. In some, diagnosis can be assisted by round-the-clock, or at least overnight, study of their secretory pattern. But the variability of 24hr-secretion rate in normal children (i.e. in children well above the 3rd centile limits for height) is so wide that some partially insufficient children can probably only be diagnosed by their growth patterns and their response to a test period of growth hormone.

The treatment of growth-hormone deficiency has been a major success of paediatrics since the first patient was given human growth hormone (HGH) by Dr Maurice Raben in 1958 in Boston. Since the hormone is species-specific growth hormone from humans had to be used and for some 25 years this was prepared from corpses at autopsy (adult's pituitaries fortunately containing as much as children's). Recently however it has been made by recombinant DNA technology, being the second human hormone to be made in this way (insulin was the first). The 'synthetic' product has the advantage also of having no risk of being contaminated by 'slow' viruses, an ever-present, if statistically remote, danger in human tissues. The hormone is given by subcutaneous injection usually every day for the duration of childhood or adolescence until growth stops naturally. Provided treatment has started at a reasonably early age (at least before age 6) the results are usually excellent (even at later ages results are sometimes spectacular, but it becomes increasingly hard to make up all the lost ground). A catch-up occurs on initial treatment and thereafter normal growth is maintained. If TSH lack is also present, thyroxine is given, and if gonadotrophin deficiency declares itself at the time of puberty (which is late in growth-hormone-deficient children, who characteristically have puberty at a normal bone age but a late chronological age), then gonadotrophin-stimulating hormone, gonadotrophins, or sex hormones have also to be supplied.

Since synthetic growth-hormone-releasing hormone (GHRH) is now available, in theory this might be another way of treating children with idiopathic growth hormone insufficiency whose failure lies in the cells of the hypothalamus and not in the pituitary itself (and this is true in the majority of cases). At the time of writing GHRH is not a practical competitor to growth hormone, however, in that it has to be given continuously or at least several times a day. If and when a long-acting analogue of GHRH is made, active in one injection a day, then the cheaper cost of GHRH compared with GH might become important.

Hypothyroidism Lack of thyroid gland secretion also stops growth occuring, for all cells seem to need a certain level of thyroid hormone in them in order to function properly. Hypothyroidism may start in utero, in which case the brain is affected and diagnosis and treatment directly after birth is a matter of urgency. The condition can be diagnosed by measuring the level of thyroid hormone or TSH in the blood, and this has been instituted as a routine screening-test for all newborns. Short stature is more usually caused by the type of hypothyroidism which starts during childhood almost insidiously. Figure 90 shows an example. The catch-up on thyroid hormone here is marked and eventually complete, as is usually the case.

6. Cartilage and Bone Disorders

There are a large number of bone disorders which affect growth: they are mostly rare and many are inherited. The best known is *achondroplasia*.

The achondroplasiac is the classical circus performer, though this is an image societies such as the UK Association for Research into Restricted Growth (ARRG) and the Little People of America (LPA) would like to expunge. Achondroplasiacs have short upper arms and thighs, a normal length back, a large head, and a characteristic face, with depressed nasal bridge, small nose and large forehead. Usually they have large muscles and are excellent athletes, whence their gravitation to circus work. They are of normal intelligence and health and usually seem to support their disorder with great cheerfulness and fortitude. Achondroplasia is due to the mutation of a single gene and is inherited as a dominant. This means that on average half the children of an achondroplasiac are affected. Most cases, however, are themselves the new mutation, without any family history: it seems that this particular gene is one of the most unstable in the human complex.

In the last few years groups of surgeons have devised leg-lengthing operations; the bony cortex of the femur is cut right round, pins inserted above and below the cut, and the distance between the pins progressively increased by external traction, by around a millimetre a day. The bone grows in to fill the artificial gap, and at least in achondroplasiacs, whose muscles, nerves and blood vessels seem to be somewhat bunched up by their specific lack of bone growth, these other tissues stretch to accomodate the extra length. Increases of leg length up to about 15 cm can be fairly reliably obtained in this way. Some surgeons operate only after growth has ceased, others before the epiphyses have fused. The arms can also be somewhat lengthened, but no treatment has been devised for the head. Growth hormone is currently being tried in a few clinics, but even if it works, the result would presumably be an increase of size unaccompanied by any change of shape, and this would be a dubious advantage.

Figure 90   Response of hypothyroid child treated with thyroxine. Solid circles, height for age; open circles, height for bone age
(From Tanner, Whitehouse, Marshall *et al.*, 1975)

A similar but less marked syndrome is hypochondroplasia. In this condition it is only the limbs that are short. The face is normal (though it has some tell-tale features to the experienced eye) so hypochondroplasiacs pass as very short normals. The disorder is also inherited as a dominant gene, but seems to be separate from the achondroplasia gene: at least the two conditions very seldom occur in the same family. In hypochondroplasia one of the parents very frequently has the disorder. There is no certain way of clinching what is sometimes a difficult diagnosis. The most helpful feature is comparison of sitting-height and leg-length centiles. While in achondroplasia the X-ray signs are marked and characteristic, in hypochondroplasia they are minimal and sometimes uncertain.

Other bone disorders are rarer: some affect limbs only, some trunk only and some both. Few are at present treatable.

For these disorders particularly, the ARRG and the LPA extend extremely valuable support. These are associations of people with short stature, together with parents of children with short stature. The members, which include doctors and surgeons with or without short stature, help one another with advice on all aspects of living, and educate the public into a better acceptance of the person handicapped in this way. Membership of the ARRG or the LPA is often the only thing a physician can offer parents of children with such bone disorders and is frequently found to be of help.

7. Diseases of Absorption of Food

If there is a prolonged malabsorption of food, growth is stunted, just as it is in starvation. All malabsorptive diseases may cause this if not treated successfully, but the most common one appearing in the growth disorder clinic is *coeliac disease*. Usually coeliac disease causes gastro-intestinal symptoms, but in a few children these are lacking and the disease is manifested solely in short stature, often but not always associated with an unusually protuberant abdomen. The diagnosis can only be made for certain by taking a biopsy of the small intestine through a tube that is swallowed. This sounds traumatic for the child but nowadays in the skilled hands of the paediatric gastroenterologist it is not so. Coeliac disease is due to an abnormal reaction of cells of the gut to substances called glutens which occur in flour and other foodstuffs. The results of treatment by a gluten-free diet are so excellent that any child with short stature not otherwise diagnosed should have a biopsy to exclude the possibility of coeliac disease.

A second gastro-intestinal disease which may (though unusually) have short stature as its first or only symptom is Crohn's disease, an inflammatory reaction of the lower gut. Diagnosis is by X-ray and specialised tests.

8. General Diseases of the Lung, Heart and Kidney

A number of chronic general diseases cause short stature, mostly for reasons that are far from clear. Kidney disease is one example, and those heart and lung diseases which cause some degree of lack of oxygen to the tissues are others. Many heart conditions, however, do not stunt the child's growth at all.

9. Steroids and Radiation

Short stature may occur as a result of treatment with steroids or X-rays, treatment which is inevitable in its proper context. Asthma, eczema, keratitis (inflammation of part of the eye) and some other conditions may if sufficiently severe yield only to treatment with cortisol or its derivatives. In children cortisol blocks the multiplication of cartilage cells and causes growth arrest. In most circumstances the effect can be minimized by giving the drug only on alternate days. No growth occurs that day, but a catch-up takes place the next, as Milner and Wales have shown using the knemometer (see p. 8). But where other drugs can be used to treat the condition cortisol should be avoided. Some cortisol compounds are readily absorbed through inflamed skin, so even their use in eczema may cause problems.

Therapeutic radiation is a necessary treatment in certain forms of leukemia. (Diagnostic X-rays are nowhere near the level that affects growth.) Irradiation of the head may cause growth hormone deficiency, but only if the dose is high; dose schedules are now worked out to minimize this effect. Radiation of the spine may cause a specific lack of spinal growth due to local damage.

10. Psychological Causes

It is now well established that psychological stress may cause short stature in certain children, who react to it by switching off their growth-hormone secretion. These are not starved children, and they are often fat, as in ordinary growth-hormone deficiency. They have a history of psychosocial stress: indeed they seem often to present an older version of the battered baby situation. One child may be the scapegoat for the whole family, just as in the battered baby syndrome, with the other children growing quite normally.

In the classical case the child has a disorder of eating; he eats voraciously at some times and not at all at others. He sleeps badly, gets up in the night and 'steals' food. Homes with children with psychosocial short stature nearly always have a lock on the door of the refrigerator. The child may eat the food set aside for the dog or cat, placing himself on the mat to do so; he rummages in dustbins and drinks water from the toilet. A stimulation test of growth-hormone secretation done immediately on arrival at the clinic shows inability to respond: the child is admitted to hospital and after a few days or weeks the response returns.

Once the diagnosis is established there is little to do but separate the child from the inimical family atmosphere (if he is sent to a school for 'delicate' children 'because he is so small' the family morale may be saved and agreement secured). The catch-up away from home is usually spectacular and fully equal to that of an idiopathically growth-hormone-lacking child given growth hormone. An example is shown in Figure 91.

Tall Stature

Though tallness is in general a prized asset in our Western industrialised society, there are limits. Some girls, and some of their (usually very tall) parents are horrified by the idea of being 180 cm (5 ft 11 in) or over and seek medical treatment to restrict their growth. Likewise, though more rarely, boys heading for 195 cm (6 ft 5 in) or so may seek a few centimetres reduction. The vast majority of tall people asking for treatment are perfectly normal and the prediction of their adult heights falls within the limits expected from their parents. Treatment with high doses of oestrogen in girls or testosterone in boys induces an early puberty and results in an average diminuation of about 5 cm in adult height. The range of effect however seems to be wide, and the prediction has a built-in error of up to 10 cm. In consequence one can never be sure in any individual case whether the treatment has worked or not: all one can say is that if a hundred children are treated, the average height they would eventually reach is some 5 cm lower than the average of their predicted heights. The earlier the treatment is instituted, of course, the greater the effect; treatment after menarche is only marginally effective, and the ideal is to start oestrogens at the earliest age at which the beginning of puberty is socially acceptable. Here again we are basically creating pathology to satisfy a real or imagined social need. The side effects of these large doses of hormones in the long term, if any, are unknown.

True gigantism so much beloved and so much illustrated by generations of medical textbooks, is in fact of the utmost rarity: few

Figure 91   Height of child with psychosocial short stature. Note catch-up on removal, marked by arrow, from parental home

practising paediatric endocrinologists have ever looked after a case. Tumours of the pituitary gland or overgrowth of the growth-hormone-producing cells certainly can cause vastly excessive growth and this is what the famous giants of old, whose skeletons adorn the medical museums, suffered from. Such malfunction is far commoner in adult life, however, when it causes acromegaly.

There are a few other syndromes which have excessive tallness as part of their spectrum. The best known is cerebral gigantism or Sotos' syndrome. The child is usually large at birth, often with particularly excessive growth of the trunk, and has a particular cast of face such that most of these patients look like brothers and sisters. The syndrome is not caused by excessive growth hormone but by an over-reaction to growth hormone by the tissues.

CODA

This chapter brings us to the end of our brief conspectus of human growth and its disorders. Provided always that parental size is known, growth emerges as a prime measure of a child's physical and mental health. The study of growth emerges also as a powerful tool for monitoring the health and nutrition of populations, especially in ecological and economic circumstances that are sub-optimal. It can be used for studying the effect of political organisation upon the relative welfare of the various social, cultural and ethnic groups which make up a modern state. Thus the study of growth has a very direct bearing upon human welfare. At the same time it provides valuable lessons on the way in which our biological heritage and our technological culture interact. It warns us that all too soon we could be technically capable of creating monstrously specialized or monstrously similar children. It points to our need to be reconciled with our origins, to see ourselves once again as a part of the natural order; not foetus into angel, nor foetus into monster, but foetus into Man.

Bibliography

Chapter 1

Further Reading

Falkner, F. and Tanner, J. M., eds (1986), *Human Growth,* 2nd edn, 3 Vols, New York and London: Plenum

The first comprehensive multi-author handbook on human growth. Chapters by M. J. R. Healy, H. Goldstein and E. Marubini respectively, are a particularly valuable introduction to biometrical methods in growth work, curve fitting and sampling for surveys. An elementary knowledge of statistics is assumed

Healy, M. J. R. (1989) 'Growth curves and growth standards: the state of the art' in *Auxology '88: Perspectives in the Science of Growth and Development,* ed. J. M. Tanner, London, Smith-Gordon; pp. 13–22

An excellent introduction to the subject of growth curve fitting

Hermanussen, M. Geiger-Benoit, K., Burmeister, J. and Sippell, W. G. (1988) 'Knemometry in childhood: accuracy and standardization of a new technique of lower leg measurement', *Annals of Human Biology,* **15,** 1–16
'Periodical changes of short-term growth velocity (mini-growth spurts) in human growth', *Annals of Human Biology,* **15,** 103–111

Two papers which describe the knemometer and provide an introduction to the subject of short-term variation in growth velocity

Marshall, W. A. (1977) *Human Growth and its Disorders,* London and New York: Academic Press

A textbook for medical graduates, which deals simply but comprehensively with human growth, chiefly from a medical point of view

Tanner, J. M.. (1962) *Growth At Adolescence,* 2nd edn, Oxford: Blackwell Scientific Publications

A textbook treatment of human growth, which despite the title, covers the whole postnatal period in some detail. Has a large bibliography still useful to the research student, though parts of the text are dated

Tanner, J. M. (1981) *A History of the Study of Human Growth,* Cambridge: University Press

A history of the subject from ancient times to the present. It aims to present both changes of thought about the subject through the centuries and all extant data concerning growth in the past two centuries

Tanner, J. M. ed. (1989) *Auxology '88: Perspectives in the Science of Growth and Development,* London: Smith-Gordon

A collection of papers from the Fifth International Congress of Auxology, held at Exeter in 1988. Twenty-two are by invited speakers who are amongst the most distinguished scholars in their fields, and a further twenty-four contributed. All the areas dealt with in this book are covered. This is the first book to which the reader who wishes to pursue the study of auxology to a deeper level should turn. It provides an unrivalled picture of contemporary research

Watts, E. S. (1986) 'The evolution of the human growth curve', in *Human Growth*, 2nd edn, Vol. 1, pp. 153–166, eds Falkner, F. and Tanner, J. M. New York: Plenum

The first good summary of our knowledge on this topic, by a prominent investigator of apes and monkeys

References

Berkey, C. S., Reed, R. B. and Valadian I. (1983), 'Mid-growth spurt in height of Boston children', *Annals of Human Biology*, **10**, 25–30

Bock, R. D. and Thissen D. (1980), 'Statistical problems of fitting individual growth curves', in *Human Physical Growth and Maturation:Methodologies and Factors*, eds Johnston, F. E., Roche, A. F. and Susanne, C. New York: Plenum

Gasser, T., Muller, H-G., Kohler, W., Prader, A., Largo, R. and Molinari, L. (1985), 'An analysis of the mid-growth and adolescent spurts of height based on acceleration', *Annals of Human Biology*, **12**, 129–148

Goss, D. A. and Winkler, R. L. (1983)'Progression of myopia in youth: age of cessation', *American Journal of Optometry and Physiological Optics*, **60**, 651–658

Hermanussen, M., Sippell, W. G. and Valk, I. M. (1985), 'Knenometric monitoring of early effects of human growth hormone on leg length in children with growth hormone deficiency', *Lancet*, **I**, 1069–1071

Israelsohn, W. J. (1960), 'Description and modes of analysis of human growth', in *Human Growth*, ed. Tanner, J. M., Oxford: Pergamon, *Symposia of the Society for the Study of Human Biology*, **3**, 21–41

Johnston, F. E., Hamill, P. V. V. and Lemeshow, S. (1974), 'Skin-fold thicknesses in a national probability sample of U.S. males and females aged 6 through 17 years', *American Journal of Physical Anthropology*, **40**, 321–324 (also D.H.E.W. Publications Nos. HSM 73–1602 and HRA 74–1614

Johnston, F. E. (1988) 'Sex differences in fat patterning in children and youth', in *Fat Distribution During Growth and Later Health Outcome*, pp. 85–102, ed. Bouchards, C. and Johnston, F. E. New York :Liss

Largo, R. H., Gasser, Th., Prader, A., Stuetzle W. and Huber, P. J. (1978), 'Analysis of the adolescent growth spurt using smoothing spline functions', *Annals of Human Biology*, **5**, 421–434

Poisonnet, C. M., Lavelle, M. and Burdi A. R. (1988) 'Growth and development of adipose tissue', *Journal of Pediatrics*, **113**, 1–9

Tanner, J.M., Whitehouse, R.H., and Takaishi, M. (1966), 'Standards from birth to maturity for height, weight, height velocity and weight velocity; British children, 1965', *Archives of Disease in Childhood*, **41**, 454–471: 613–635

Tanner, J. M. and Whitehouse, R. H. (1975), 'Revised standards for triceps and subscapular skinfolds in British children', *Archives of Disease in Childhood*, **50**, 142–145

Tanner, J. M., and Whitehouse, R. H. (1976), 'Clinical longitudinal standards for height, weight, height velocity, weight velocity and the stages of puberty', *Archives of Disease in Childhood*, **51**, 170–179

Chapter 2

Further Reading

Goss, R. J. (1966), 'Hypertrophy versus hyperplasia', *Science*, **153**, 1615–1620

A good short introduction to cellular growth

Goss, R. (1986), 'Modes of growth and regeneration: mechanisms, regulation, distribution', in *Human Growth*, 2nd edn, Vol 1, pp. 1–26, eds Falkner, F. and Tanner, J. M. , New York: Plenum

A general and very readable account of tissue growth, with a description of experiments demonstrating some of the mechanisms of its control

Isaksson, O. G. P., Lindahl, A., Nilsson, A. and Isgaard, J. (1987) 'Mechanism of the stimulatory effect of growth hormone on longitudinal bone growth', *Endocrine Reviews*, **8**, 426–438

An excellent and simply written review of the physiology of the growth plate by the group which in the 1980s transformed the understanding of its endocrinology

Jacobson, M. (1974), 'Differentiation and growth of nerve cells', in *Differentiation and Growth of Cells in Vertebrate Tissues*, ed. Goldspink, G., London: Chapman and Hall

A clear and exellent resume of the subject, suitable, however, only for readers with some knowledge of biology

Winick, M. (1971), 'Cellular growth during early malnutrition', *Pediatrics*, **47**, 969–978

A general account of the author's work on undernutrition and the brain in animals. Defends theory that the more rapid the rate of cell division at the time of starvation the more lasting the effect

Widdowson, E. M. (1970), 'Harmony of Growth', *Lancet*, **I**, 1901–1905

A lecture, to a general audience, which provides a good starting-point for reading about comparative growth in the rat, pig, man, whale and other mammals. Emphasis is on cellular and tissue growth

References

Brasel, J. A. (1974), 'Cellular changes in intra-uterine malnutrition', in *Nutrition and Fetal Development*, ed. Winick, M., New York: Wiley

Cheek, D. B. (1968), 'Muscle cell growth in normal children', in *Human Growth*, ed. Cheek, D. B., Philadelphia: Lea and Febiger

Goldspink, G. (1974), 'Development of muscle', in *Differentiation and Growth of Cells in Vertebrate Tissues*, ed. Goldspink, G., London: Chapman and Hall

Goss, R. J. (1981), 'Physiological adaptations of growth', *Comparative Animal Nutrition*, **4**, 1–32

Hunziker, E. B. (1988) 'Growth plate structure and function', *Pathology and Immunopathology Research* **7**, 9–13

Jacobson, M. (1970), *Developmental Neurobiology*, New York: Holt, Rhinehart and Winston

Johnston, F. E., Weston, M., McKinney, S., Coleman, J., Pereira, G. and Rounds, J. (1984), 'Differences in the mean fat cell diameter of males between 1 and 48 months of age', *American Journal of Physical Anthropology*, **65**, 341–346

Kember, N. F. and Sissons, H. A. (1976), 'Quantitative history of the human growth plate', *Journal of Bone and Joint Surgery*, **58-B**, 426–435

Pritchard, J. J. (1974), 'Growth and differentiation of bone and connective tissue', in *Differentiation and Growth of Cells in Vertebrate Tissues*, ed. Goldspink, G., London: Chapman and Hall

Steffe, W. P., Goldsmith, R. S., Pencharz, P. B., Scrimshaw, N. S. and Young, V. R. (1976), 'Dietary protein intake and dynamic aspects of whole body nitrogen metabolism in adult humans', *Metabolism*, **25**, 281–297

Thorngren, K.-G., and Hansson, L. I. (1981), 'Cell production of different growth plates in the rabbit', *Acta Anatomica*, **110**, 121–127

Waterlow, J. C. (1968), 'Observations on the mechanisms of adaptation to low protein intake', *Lancet*, **II**, 1091–1094

Waterlow, J. C. (1975), 'Protein turnover in the whole body', *Nature*, **253**, 157

Zezulak, K. M. and Green, H. (1986) 'The generation of insulin-like growth factor I – sensitive cells by growth hormone action', *Science*, **233**, 551–553

Chapter 3

Further Reading

Alberman, E. and Creasy, M. R. (1975), 'Factors affecting chromosome abnormalities in human conceptions', in *Chromosome Variations in Human Evolution*, ed. Boyce, A. J., *Symposia of the Society for the Study of Human Biology*, **14**, 83–96

A review of foetal mortality due to chromosome abnormalities

Deter, R. L., Harrist, R. B., Hadlock, F. P. and Poindexter, A. N. (1982) 'Longitudinal studies of fetal growth with the use of dynamic image ultrasonography', *American Journal of Obstetrics and Gynecology*, **143**, 545–554

An account of the outstanding longitudinal study by ultrasound made to date

McLaren, A. (1968), 'Life before birth: the first few days', *Proceedings of the Royal Institution*, **42**, 153–169

A Royal Institution lecture, before a lay audience. Discusses embryonic development, differentiation in the embryo, and the techniques and results of egg transfer from one mother to another, in which subject the author was a pioneer

Meire, H. B. (1986), 'Ultrasound measurement of fetal growth', in *Human Growth*, 2nd edn, Vol I, eds Falkner, F. E. and Tanner, J. M., New York: Plenum

An up-to-date review of the contributions of ultrasound to knowledge of foetal growth

Neilson, L., Ingleman-Sunberg, A. and Winsen, C. (1967), *The Everyday Miracle: A Child is Born*, London: Lane, Penguin Press

A popular account of the growth of a child from conception to birth, distinguished by Neilson's beautiful pictures of human embryos and foetuses, which allow the reader to visualize exactly the changes of form, the situation in the uterus and the development of head, hands, feet and face

Tanner, J. M. (1974), 'Variability of growth and maturity in newborn infants', in *The Effect of the Infant on its Caregiver*, eds Lewis, M. and Rosenblum, L. A., New York: Wiley

An account which stresses the extent of individual variability in size and maturity, even at birth

Walton, A. and Hammond, J. (1938), 'The maternal effects on growth and conformation in Shire horse–Shetland pony crosses', *Proceedings of the Royal Society B*, **125**, 311–335

The classic experiment of crossing Shire with Shetland horses, to observe the size of the foal at birth and its subsequent growth

References

Birkbeck, J. A. (1976), 'Metrical growth and skeletal development of the human fetus', in *The Biology of Fetal Growth*, eds Roberts, D. F. and Thomson, A. M., *Symposia of the Society for the Study of Human Biology*, **15**, 39–68

Brandt, I. (1976), 'Dynamics of head circumference growth before and after term', in *The Biology of Fetal Growth*, eds Roberts, D. F. and Thomson, A. M., *Symposia of the Society for the Study of Human Biology*, **15**, 109–136

Brenner, W. E., Edelman, D. A. and Hendricks, C. H. (1976), 'A standard of fetal growth for the United States of America', *American Journal of Obstetrics and Gynecology*, **126**, 555–564

Briend, A. (1980), 'Maternal physical activity, birth weight and perinatal mortality', *Medical Hypotheses*, **6**, 1157–1170

Butler, N. R., Goldstein, H. and Ross, E. M. (1972), 'Cigarette smoking in pregnancy: its influence on birth weight and perinatal mortality', *British Medical Journal*, **1**, 127

Dargassies, S. St-A. (1966), 'Neurological maturation of the premature infant of 28 to 41 weeks gestational age', in *Human Development*, ed. Falkner, F., London: Saunders

Deter, R. L., Rossavik, I. V., Harrist, R. B. and Hadlock, F. P. (1986), 'Mathematic modeling of fetal growth:development of individual growth curve standards', *Obstetrics and Gynecology*, **68**, 156–161

Dreyfus-Brisac, C. (1975), 'Neurophysiological studies in human premature and full-term newborns', *Biological Psychiatry*, **10**, 485–496

Drillien, C. M. (1964), *The Growth and Development of the Prematurely Born Infant*, Edinburgh: Livingstone

Fancourt, R., Campbell, S., Harvey, D. and Norman, A. P. (1976), 'Follow-up study of small-for-dates babies', *British Medical Journal*, **1**, 1435–1437

Goeschen, K., Pluta, M., Rothe, J. and Saling, E. (1983), 'Motor nerve conduction velocity for assessment of gestational age in twins', *Gynecological and Obstetric Investigations*, **15**, 74–83

Jakobovits, A., Iffy, L., Wingate, M. B., Slate, W. G., Chatterton, R. T. and Kerner, P. (1972), 'The rate of early fetal growth in the human subject', *Acta Anatomica*, **83**, 50–59

Jones, K. L. and Smith, D. W. (1975), 'The fetal alcohol syndrome', *Teratology*, **12**, 1–10

Keen, D. V. and Pearse, R. G. (1988), 'Weight, length, and head circumference curves for boys and girls of between 20 and 42 week's gestation', *Archives of Disease in Childhood*, **63**, 1170–1172

Knobloch, H. and Pasamanick, B. (1963), 'Mental subnormality', *New England Journal of Medicine*, **266**, 1045–1051; 1092–1097; 1155–1161

Lechtig, A., Delgado, H., Lasky, R. E., Klein, R. E., Engles, P. L., Yarbrough, C. and Habicht, J. P. (1975) 'Maternal nutrition and fetal growth in developing societies', *American Journal of Diseases of Children*, **129**, 434–437

Moosa, A. and Dubowitz, V. (1971), 'Postnatal maturation of peripheral nerves in pre-term and full-term infants', *Journal of Pediatrics*, **79**, 915–922

Nilsen, S. T., Finne, P. H., Bergsjø, P. and Stamnes, O. (1984), 'Males with low birthweight examined at 18 years of age', *Acta Paedatrica Scandinavica*, **73**, 168–175

Poissonnet, C. M., Burdi, A. R. and Garn, S. M. (1984), 'The chronology of adipose tissue appearance and distribution in the human fetus', *Early Human Development*, **10**, 1–11

Smith, D. W. (1977), *Growth and its Disorders*, Philadelphia: Saunders

Southgate, D. A. T. and Hay, E. N. (1975), 'Chemical and biochemical development of the fetus', in *The Biology of Fetal Growth*, eds Roberts, D. F. and Thomson, A. M., *Symposia of the Society for the Study of Human Biology*, **15**, 195–209

Tanner, J. M. (1963), 'The regulation of human growth', *Child Development*, **34**, 817–847

Tanner, J. M. and Thomson, A. M. (1970), 'Standards for birth-weight at gestation periods from 32 to 42 weeks allowing for maternal height and weight', *Archives of Disease in Childhood*, **45**, 566–569

Usher, R. H. and McLean, F. H. (1974), 'Normal fetal growth and the significance of fetal growth retardation', in *Scientific Foundations of Paediatrics*, eds Davis, J. A. and Dobbing, J., London: Heinemann

Yudkin, R. I., Aboualfa, M., Eyre, I. A., Redman, C. W. G. and Wilkinson, A. R. (1987), 'Influence of elective preterm delivery on birthweight and head circumference standards', *Archives of Disease in Childhood*, **62**, 24–29

Chapter 4

Further Reading

McEwen, B. S. (1976), 'Interactions between hormones and nerve tissues', *Scientific American*, **235**, 48–58

An account of the way in which sex hormones cause sex differentiation of the brain. Also discusses oestradiol and corticosterone receptors in brain. Very well illustrated

Money, J. and Ehrhardt, A. A. (1972), *Man and Woman: Boy and Girl*, Baltimore: Johns Hopkins University Press

Already a classic: written from the viewpoint of psychologists dealing with patients with intersexual and gender identity problems. The early chapters discuss sex differentiation of body structure and the later ones differentiation of gender identity and gender role. One of the very few books in which a satisfactory balance is held between the endocrinological and psycho-social aspects of sexual differentiation

Ratcliffe, S. C. (1975), 'Postnatal chromosomal abnormalities', in *Chromosome Variations in Human Evolution*, ed. Boyce, A. J., *Symposia of the Society for the Study of Human Biology*, **14**, 97–115

Gives worldwide data on incidence of chromosomal abnormalities at birth: some 43,000 presumed normal newborns had had karyotypes done at this time of reporting

References

Alberman, E. and Creasy, M. R. (1975), 'Factors affecting chromosome abnormalities in human conceptions', in *Chromosome Variations in Human Evolution*, ed. Boyce, A. J., *Symposia of the Society for the Study of Human Biology*, **14**, 83–95

Eayrs, J. T. (1971), 'Thyroid and developing brain: anatomical and behavioural effects', in *Hormones in Development*, eds Hambrugh, M. and Barrington, E. J. W., New York: Appleton–Century

Feldman, K. W. and Smith, D. W. (1975), 'Fetal phallic growth and penile standards for newborn male infants', *Journal of Pediatrics*, **86**, 395–398

George, F. W., Griffin, J. E., Leshin, M. and Wilson, J. D. (1981), 'Endocrine control of sexual differentiation in the human', in *Fetal Endocrinology*, eds Novy, M. J. and Ruskò, J. A., pp. 341–357, New York: Academic Press

Josso, N. (1974), 'Mullerian inhibiting activity of human fetal testicular tissue deprived of germ cells by in vitro irradiation', *Pediatric Research*, **8**, 755–758

Kuhns, L. R. and Finnstrøm, O. (1976), 'New standards of ossification of the newborn', *Radiology*, **119**, 655–660

Reyes, F. I., Winter, J. S. D. and Faiman, C. (1973), 'Studies on human sexual development. 1. Fetal gonadal and adrenal sex steroids', *Journal of Clinical Endocrinology and Metabolism*, **37**, 74–78

Tanner, J. M., Prader, A., Habicht, H., and Ferguson-Smith, M. A. (1959), 'Genes on the Y chromosome influencing rate of development in Man. Skeletal age studies in Klinefelter's (XXY) and Turner's (XO) syndromes', *Lancet*, (ii), 141–144

Chapter 5

Further Reading

Billewicz, W. Z., Fellowes, H. M. and Thomson, A. M. (1981), 'Pubertal changes in boys and girls in Newcastle-upon-Tyne', *Annals of Human Biology*, **8**, 211–219

A comprehensive report on an important longitudinal study of puberty in Newcastle-upon-Tyne

Marshall, W. A. and Tanner, J. M. (1969), 'Variations in the pattern of pubertal changes in girls', *Archives of Disease in Childhood*, **44**, 291–303

Marshall, W. A. and Tanner, J. M. (1970), 'Variations in the pattern of pubertal changes in boys', *Archives of disease in Childhood*, **45**, 13–23

Two papers describing pubertal variations between individuals, based on the largest number of children yet followed longitudinally and in detail throughout puberty

Prader, A. (1984), 'Biometrical and endocrinological aspects of normal growth and development', in *Human Growth and Development*, eds Borms, J., Hauspie, R., Sand, A., Susanne, C. and Hebbelinck, M., pp. 1–22, New York: Plenum

An excellent presentation of the results of the First Zurich Longitudinal Growth Study, 1954–80, with particular emphasis on the morphology and endocrinology of puberty

Tanner, J. M. (1962), *Growth at Adolescence*, 2nd edn, Oxford: Blackwell Scientific Publications

Covers pubertal changes in some detail, with extensive bibliography

Tanner, J. M. (1975), 'Growth and endocrinology of the adolescent', in *Endocrine and Genetic Diseases of Childhood*, 2nd edn, ed. Gardner, L., Philadelphia and London: Saunders

An account of pubertal changes, and their endocrinological causes, suitable for biologists as well as doctors

Taranger, J., Engström, I., Lichtenstein, H. and Svennberg-Redegren, I. (1976), 'The somatic development of children in a Swedish urban community. VI. Somatic pubertal development', *Acta Paediatrica Scandinavica*, Supplement, 258, 121 135

A report of the Stockholm Longitudinal Growth Study (parallel with studies in Brussels, London, Paris and Zurich) giving details in particular of ages of transition from each stage of puberty to the next

References

Billewicz, W. Z., Fellowes, H. M. and Thomson, A. M. (1980), 'Post-menarcheal menstrual cycles in British (Newcastle-upon-Tyne) girls', *Annals of Human Biology*, **7**, 177–180

Bielicki, T., Koniarek, J. and Malina R. M. (1984), 'Inter-relationships among certain measures of growth and maturation rate in boys during adolescence', *Annals of Human Biology*, **11**, 201–210

Burke, P. H. and Hughes-Lawson, C. A. (1988), 'The adolescent growth spurt in the soft tissues of the face', *Annals of Human Biology*, **15**, 253–262

Cronk, C. E., Roche, A. F., Kent, R. Jr., Eichorn D. and McCammon, R. W. (1983), 'Longitudinal trends in subcutaneous fat thickness during adolescence', *American Journal of Physical Anthropology*, **61**, 197–203

Haas, J. H. (1971), *Growth Diagrams 1965, Netherlands*, Leiden: Institute of Preventive Medicine

Kemper, H. C. G. and Verschuur, R. (1987), 'Longitudinal study of maximal aerobic power in teenagers', *Annals of Human Biology*, **14**, 435–444

Jones, H. E. (1949), *Motor Performance and Growth. A Developmental Study of Static Dynamometric Strength*, Berkeley: University of California Press

Largo, R. H. and Prader, A. (1983a), 'Pubertal development in Swiss boys', *Helvetica Paediatrica Acta*, **38**, 211–228

Largo, R. H. and Prader, A. (1983b), 'Pubertal development in Swiss girls', *Helvetica Paediatrica Acta*, **38**, 229–243

Lindgren, G. (1978), 'Growth of schoolchildren with early, average and late ages of peak height velocity', *Annals of Human Biology*, **5**, 253–267

Marshall, W. A. and Tanner, J. M. (1986), 'Puberty', in *Human Growth*, 2nd edn, Vol 2, eds Falkner, F. and Tanner, J. M., New York: Plenum

Metcalf, M. G., Skidmore, D. S., Lowry, G. F. and Mackenzie, J. A. (1983), 'Incidence of ovulation in the years after the menarche', *Journal of Endocrinology*, **97**, 213–219

Mirwald, R. L., Bailey, D. A., Cameron, N. and Rasmussen, R. L. (1981), 'Longitudinal comparison of aerobic power in active and inactive boys aged 7.0 to 17.0 years', *Annals of Human Biology*, **8**, 405–414

Nielsen, C. T., Skakkebaek, N. E., Richardson, D. W., Darling, J. A. B., Hunter, W. M., Jergensen, M., Nielsen, A., Ingerslev, O., Keiding, N. and Muller, J. (1986), 'Onset of the release of spermatazoa (spermarche) in boys in relation to age, testicular growth, pubic hair and height', *Journal of Clinical Endocrinology and Metabolism*, **62**, 532–535

Rutenfranz, J., Andersen, K. L., Seliger, V., Ilmarinen, J., Klimmer, F., Kylian, H., Rutenfranz, M. and Ruppel, M. (1982), 'Maximal aerobic power affected by maturation and body growth during childhood and adolescence', *European Journal of Pediatrics*, **139**, 106–112

Salardi, S., Orsini, L. F., Cacciari, E., Bovicelli, L., Tassoni, P. and Reggiani, A. (1985), 'Pelvic ultrasonography in pre-menarcheal girls: relation to puberty and sex hormone concentrations', *Archives of Disease in Childhood*, **60**, 120–125

Stolz, H. R. and Stolz, L. M. (1951), *Somatic Development of Adolescent Boys. A Study of the Growth of Boys during the Second Decade of Life*, New York: Macmillan

Tanner, J. M., Hughes, P. C. R. and Whitehouse, R. H. (1981), 'Radiographically determined widths of bone muscle and fat in upper arm and calf from age 3 to 18 years', *Annals of Human Biology*, **8**, 495–518

Zacharias, L. and Rand, W. M. (1983), 'Adolescent growth in height and its relation to menarche in contemporary American girls', *Annals of Human Biology*, **10**, 209–222

Zachmann, M., Prader, A., Kind, H. P., Haflinger, H. and Budliger, H. (1974), 'Testicular volume during adolescence', *Helvetica Paediatrica Acta*, **29**, 61–72

Chapter 6

Further Reading

Acheson, R. M. (1966), 'Maturation of the skeleton', in *Human Development*, ed. Falkner, F., Philadelphia: Saunders

An excellent introduction to the assessment of skeletal maturity by a pioneer of the scoring method

Marshall, W. A. and Limongi, Y. (1976), 'Skeletal maturity and the prediction of age at menarche', *Annals of Human Biology*, **3**, 235–243

An important paper, in method as well as result. Requires only elementary knowledge of statistics, none of biology

Prokopec, M. (1982), 'Early and late maturers', *Anthrop. Kozl.*, **26**, 13–24

An English-language article describing the growth of a group of early - and a group of late-maturers, in a longitudinal study in Prague, with remarks on psychological attitudes

Tanner, J. M. (1966), 'Galtonian eugenics and the study of growth. The relation of body size, intelligence test score and social circumstances in children and adults', *Eugenics Review*, **58**, 122–135, reprinted in *Trends and Issues in Developmental Psychology*, eds Mussen, P. H., Largen, J. and Covington, M., New York: Holt, 1969

A review, citing the historical background as well as modern studies

Tanner, J. M., Whitehouse, R. H., Cameron, N., Marshall, W. A., Healy, M. J. R. and Goldstein, H. (1983), *Assessment of Skeletal Maturity and Prediction of Adult Height*, 2nd edn, London: Academic Press

A short monograph describing the Tanner–Whitehouse system of bone age and giving predictions of adult height based upon it. The early chapters discuss the general context of developmental age assessment

Tanner, J, M (1978), *Education and Physical Growth*, 2nd edn, London: Hodder and Stoughton

A textbook for teachers. Contains more detail on the matters discussed in this chapter and Chapter 9, together with implications for education

References

Baughan, B., Demirjian, A. and Levesque, G. Y. (1979), 'Skeletal maturity standards for French-Canadian children of school-age with a discussion of the reliability and validity of such measures', *Human Biology*, **51**, 353–370

Demerjian, A., Goldstein, H. and Tanner, J. M.(1973), 'A new system of dental age assessment', *Human Biology*, **45**, 211–227

Douglas, J. W. B. and Ross, J. M. (1964), 'Age of puberty related to educational ability, attainment and school leaving age', *Journal of Child Psychology and Psychiatry*, **5**, 185–196

Garn, S. M., LaVelle, M., Rosenburg, K. R. and Hawthorne, V. M. (1986), 'Maturational timing as a factor in female fatness and obesity', *American Journal of Clinical Nutrition*, **43**, 879–883

Greulich, W. W. and Pyle, S. I. (1959) *Radiographic Atlas of Skeletal Development of the Hand and Wrist*, 2nd edn, Stanford, California: Stanford University Press

Linos, A., Gray, J. E., Orvis, A, L., Kyle, R. A., O'Fallon, W. M. and Kurland, I, T. (1980), 'Low-dose radiation and leukemia', *New England Journal of Medicine*, **302**, 1101–1105

Marshall, W. A. (1974), 'Interrelationships of skeletal maturation, sexual development and somatic growth in man', *Annals of Human Biology*, **1**, 29–40

Mussen, P. H. and Jones, M. C. (1957), 'Self-conceptions, motivations and interpersonal attitudes of late- and early-maturing boys', *Child Development*, **28**, 243–256

Mussen, P. H. and Jones, M. C. (1958), 'The behaviour-inferred motivations of late- and early-maturing boys', *Child Development*, **29**, 61–67

Notellmann, E. D., Susman, E. J., Blue, J. H., Inoff-Germain, G., Dorn, L. D., Loriaux, D. L., Cutler, G. B. Jr. and Chrousos, G. P. (1986), 'Gonadal and adrenal hormone correlates of adjustment in early adolescence', in *Biological-Psychosocial Interactions in Early Adolescence: A Life-Span Perspective*, eds Lerner, R. M. and Foch, T. T., Hillsdale, NJ: Erlbaum

Roche, A. F., Wainer, H. and Thissen, D. (1975), *Skeletal Maturity: The Knee as a Biological Indicator*, New York: Plenum

Taranger, J. and Karlberg, P. (1976), 'The somatic development of children in a Swedish urban community. VII. Graphic analysis of biological maturation by means of maturograms', *Acta Paediatrica Scandinavica*, supplement, **258**, 136–146

Chapter 7

Further Reading

Kelch, R. P., Marshall, J. C., Sauder, S., Hopwood, N. J. and Reame, N. E. (1983), 'Gonadotropin regulation during human puberty', in *Neuroendocrine Aspects of Reproduction*, ed., Reid, R. L., New York: Academic Press

A review of pubertal endocrinology in man by one of the principle groups of investigators

Crowley, W. F. and Hoffler, J. G. eds (1987) *The Episodic Secretion of Hormones*, New York: Wiley

A collection of thirty-three papers given in 1985 covering comprehensively the detection, occurrence and importance of hormone pulses. Gives an excellent insight into ongoing research

Knobil, E. (1980), 'The neuroendocrine control of the menstrual cycle', *Recent Progress in Hormone Research*, **36**, 53–88

A very clear review of his work by the leading master in the field

O'Malley, B. W. and Schrader, W. T. (1976), 'The receptors of steroid hormones', *Scientific American*, **234**, 32–43

A clear and detailed account of how cytoplasmic hormone receptors work, by two of the chief investigators in this field. Strongly recommended to readers with some knowledge of biology

Plant, T. M. (1986), 'Gonadal regulation of hypothalamic gonadotropin-releasing hormone release in primates', *Endocrine Reviews*, **7**, 75–88

A very well written review of experiments, many by Plant himself, showing the existence of the 'sleeping phase' of the hypothalamus in the male primate

Preece, M. A. (1986) 'Prepubertal and pubertal endocrinology', in *Human Growth*, 2nd edn, Vol 2, eds Falkner, F. and Tanner, J. M., New York: Plenum

A summary of the endocrine events of puberty based on Preece's own longitudinal study

Tannenbaum, G. S. (1988) 'Somatostation as a physiological regulator of pulsatile growth hormone secretion', *Hormone Research*, **29**, 70–74

An excellent review, with good illustrations, of the control of GH secretion by hypothalamic hormones

Wierman, M. E. and Crowley, W. F. Jr. (1986), 'Neuroendocrine control of the onset of puberty', in *Human Growth*, 2nd edn, Vol 2, eds Falkner, F. and Tanner, J. M. New York: Plenum

A concise review of the endocrine control of puberty

References

Binoux, M., Gourmelen, M. and Girard, F. (1986), 'Serum levels of insulin-like growth factor (IGF) and IGF-binding protein in constitutionally short children and adolescents', *Acta Endocrinologica*, **113**, 145–152

Cassorla, F. G., Golden, S. M., Johnsonbaugh, R. E., Heroman, W. M., Loriaux, D. L. and Sherins, R. J. (1981), 'Testicular volume during early infancy', *Journal of Pediatrics*, **99**,742–743

Clark, R. G. and Robinson, I, C, A F (1985), 'Growth hormone responses to multiple injections of a fragment of human growth hormone-releasing factor in conscious male and female rats', *Journal of Endocrinology*, **106**, 281–289

Conte, F. A., Grumbach, M. M., Kaplan, S. L. and Reiter, E. O. (1980), 'Correlation of luteinizing hormone-releasing factor-induced luteinizing hormone and follicle-stimulating hormone release from infancy to 19 years with the changing pattern of gonadotropin secretion in agonadal patients – relation to the restraint of puberty', *Journal of Clinical Endocrinology and Metabolism*, **50**, 163–168

Crowley, W. F. Jr., Filicori, M., Spratt, D. I. and Santoro, N. F. (1985), 'The physiology of gonadotropin-releasing hormone (GnRH) secretion in men and women', *Recent Progress in Hormone Research*, **41**, 473–531

Cuttler, L., Van Vliet, G., Conte, F. A., Kaplan, S. L. and Grumbach, M. M. (1985), 'Somatomedin-C levels in children and adolescents with gonadal dysgenesis: differences from age-matched normal females and effect of chronic estrogen replacement therapy', *Journal of Clinical Endocrinology and Metabolism*, **60**, 1087–1092

Forest, M. G., Peretti, E. de., David, M. and Sempé, M. (1982), 'Does the adrenarche play a role in pubertal development? Evidence against it derived from the study of six clinical situations of dissociation between adrenarche and gonadarche and from the failure of a long-term treatment with dehydroepiandrosterone sulphate of children with delayed adrenarche', *Annales d'Endocrinologie*, **43**, 465–496

Gupta, D. (1986), 'The pineal gland in relation to growth and development in children', *Journal of Neural Transmission*, **21**, 217–232

Hindmarsh, P. C., Matthews, D. R. and Brook, C. G. D. (1988) 'Growth hormone secretion in children determined by time series analysis', *Clinical Endocrinology* **29**, 35–44

Ho, K. Y., Evans, W. S., Blizzard, R. M., Veldhuis, J. D., Marriam, G. R., Samojlik, E., Furlanetto, R., Rogol, A. D., Kaiser, D. L. and Thorner, M. O. (1987), 'Effects of sex and age on the 24 hour secretory profile of GH secretion in man: importance of endogenous estradiol concentrations', *Journal of Clinical Endocrinology and Metabolism*, **64**, 51–58

Ho, K. Y., Veldhuis, J. D., Johnson, M. L., Furlanetto, R., Evans, W. S., Alberti, K. G. M. and Thorner, M. O. (1988) Fasting enhances growth hormone secretion and amplifies the complex rythms of growth hormone secretion in Man', *Journal of Clinical Investigation*, **81**, 968–975

Isley, W. L., Underwood, L. E. and Clemmons, D. R. (1983), 'Dietary components that regulate serum somatomedin-C concentrations in humans', *Journal of Clinical Investigation*, **71**, 175–182

Jansson, J. O., Eden, S. and Isaksson, O. (1985), 'Sexual dimorphism in the control of growth hormone secretion', *Endocrine Reviews*, **6**, 128–150

Kelch, R. P., Khoury, S. A., Hale, P. M., Hopwood, N. J. and Marshall, J. C. (1987) 'Pulsatile secretion of gonadotrophins in children', in *The Episodic Secretion of Hormones*, eds Crowley, W. F. and Hofler, J. G. pp. 187–199, New York: Wiley

Link, K., Blizzard, R. M., Evans, W. S., Kaiser, D. L., Parker, M. W. and Rogol, A. D. (1986), 'The effect of androgens on the pulsatile release and the 24-hour mean concentration of growth hormone in peripubertal males', *Journal of Clinical Endocrinology and Metabolism*, **62**, 159–164

McCusker, R. H. and Campion, D. R. (1986), 'Effect of growth hormone-secreting tumours on skeletal muscle cellularity in the rat', *Journal of Endocrinology*, **111**, 279–285

Nilsson, A., Isgaard, J., Lindahl, A., Dahlstrom, A., Skottner, A. and Isaksson, O. G. P. (1986), 'Regulation by growth hormone of numbers of chondrocytes containing IGF in rat growth plate', *Science*, **233**, 571–574

Plant, T. M. (1985), 'A study of the role of the postnatal testes in determining the ontogeny of gonadotropin secretion in the male Rhesus monkey *(Macaca mulatta)*', *Endocrinology*, **116**, 1341–1350

Plant, T. M. and Zorub, D. S. (1984), 'A study of the role of the adrenal glands in the initiation of the hiatus in gonadotropin secretion during prepubertal development in the male Rhesus monkey *(Macaca mulatta)*', *Endocrinology*, **114**, 560–565

Prader, A., Tanner, J. M. and von Harnack, G. A. (1963), 'Catch-up growth following illness or starvation', *Journal of Pediatrics*, **62**, 646–659

Rapisdara, J. J., Bergman, K. S., Steiner, R. A. and Foster, D. L. (1983), 'Response to estradiol inhibition of tonic luteinizing hormone secretion decreases during the final stage of puberty in the Rhesus monkey', *Endocrinology*, **112**, 1172–1179

Stanhope, R., Brook, C. G. D., Pringle, P. J., Adams, J. and Jacobs, H. S. (1987), 'Induction of puberty by pulsatile gonadotrophin releasing hormone', *Lancet* **(ii)** 552–555

Stanhope, R., Pringle, P. J. and Brook, C. G. D. (1988) 'The mechanism of the adolescent growth spurt induced by low pulsatile GnRH treatment', *Clinical Endocrinology*, **28**, 83–92

Takano, K., Honda, N., Shizume, K., Hizuka, N. and Ling, N. C. (1984), 'Plasma GH and somatomedin-C responses to continuous growth hormone-releasing factor infusion in normal adult men', *Endocrinologica Japonica*, **32**, 287–293

Veldhuis, J. D., Evans, W. S., Johnson, M. L., Wills, M. R. and Rogol, A. D. (1986), 'Physiological properties of the luteinizing hormone pulse signal: impact of intensive and extended venous sampling paradigms on its characterization in healthy men and women', *Journal of Clinical Endocrinology and Metabolosm*, **62**, 881–891

Wales, J. K. H. and Milner, R. D. G. (1988) 'Variation in lower leg growth with alternate day steroid treatment', *Archives of Disease in Childhood*, **63**, 981–983

Wildt, L., Marshall, G. and Knobil, E. (1980), 'Experimental induction of puberty in the infantile female Rhesus monkey', *Science*, **207**, 1373–1374

Chapter 8

Further Reading

Becker, L. E., Armstrong, D. L., Chan, F. and Wood, M. M. (1984), 'Dendritic development in human occipital cortical neurons', *Developmental Brain Research,* **13**, 117–124

The best of the very few quantitative studies on growth of dendrites in the human: (occipital cortex, from 20 weeks' gestation to 7 years). Clearly and simply presented

Brandt, I. (1976), 'Dynamics of head circumference growth before and after term', in *The Biology of Human Fetal Growth,* eds Roberts, D. F. and Thomson, A. M., *Symposia of the Society for the Study of Human Biology,* **15**, 109–136

A review of the growth of head circumference, incorporating the results of the author's own very thorough longitudinal study of pre-term infants. Highlights difficulties of obtaining accurate information on head and brain growth

Changeux, J. -P. and Dauchin, A. (1976), 'Selective stabilisation of developing synapses as a mechanism for the specification of neuronal networks', *Nature,* **264**, 705–712

An excellent review, though suitable only for readers with a knowledge of physiology and biochemistry

Gall, C., Gwen, I. and Lynch, G. (1986), 'Neuroanatomical plasticity: its role in organizing and re-organizing the central nervous system', in *Human Growth,* 2nd edn, Vol 2, eds Falkner, F. and Tanner, J. M., New York: Plenum

A review summarizing the evidence that the brain and in particular the immature brain, reacts to an excess or a deprivation of particular stimuli by altering the size and structure of neurons and neuroglia

Gaze, R. M. and Keating, J. M. eds (1974), 'Development and regeneration in the nervous system', *British Medical Bulletin,* **30**, 105–189

Fifteen essays ranging from the basis of neuronal specificity to the development of infant behaviour. Biological or medical knowledge is assumed, but no special acquaintance with the nervous system

Jacobson, M. and Hunt, R. K. (1973), 'The origins of nerve-cell specificity', *Scientifc American,* **228**, 26–35

Not an easy article, but gives a real insight into the painstaking world of experimental embryology. The authors describe their experiments on the mechanism by which nerve cells from the retina achieve the exact point-to-point projection on the brain centres subserving vision. A feel for three-dimensional structures is helpful

Konishi, M. and Akutagawa, E. (1985), 'Neuronal growth, atrophy and death in a sexually dimorphic song nucleus in the zebra finch brain', *Nature,* **315**, 145–147

A brief but cogent description of the hormone-controlled differentiation of the song nucleus in the brain of the finch

Llinas, R. R. (1975), 'The cortex of the cerebellum', *Scientific American*, **232**, 56–71

A good introduction to the relation between structure and function in the central nervous system. The cerebellum is better understood than other parts of the brain, and its relatively simple structure is beautifully illustrated in this article. No knowledge of physiology is assumed; straightforward to anyone used to wiring diagrams or blue-prints

Lynn, M. P., Atkinson, M. B. and Breuer, A. C. (1986) 'Influence of translocation track on the motion of intra-axonally transported organelles in human nerve', *Cell Motility and the Cytoskeleton*, **6**, 339–346

A description of the tracks found in axons, with organelles passing up and down from nerve cell body to periphery

Mark, R. (1974), *Memory and Nerve Cell Connections*, Oxford: Clarendon
A most stimulating introduction to the nervous system from the point of view of its development and organization. Though authoritative, it is not overburdened with technology, and can be understood by any reader who has an elementary idea of electronics or biology. For the more advanced, still an excellent introduction to neurophysiology

Sidman, R. L. and Rakic, P. (1974), 'Neuronal migrations in human brain development', *Modern Problems in Paediatrics*, **23**, 13–43

A highly technical account, but with beautiful reconstructions, and full of ideas

References

Blakemore, C. (1976), 'The conditions required for the maintenance of binocularity in the kitten's visual cortex', *Journal of Physiology*, **261**, 423–444

Burn, J., Birkbeck, J. A. and Roberts, D. F. (1975), 'Early fetal brain growth', *Human Biology*, **47**, 511–522

Cheek, D. B. (1975), *Fetal and Postnatal Cellular Growth*, New York: Wiley

Cheek, D. B., Holt, A. B., London, W. T., Ellenberg, J. H., Hill, D. E. and Sever, J. L. (1976), 'Nutritional studies in the pregnant rhesus monkey: the effect of protein-calorie or protein deprivation on growth of the fetal brain', *American Journal of Clinical Nutrition*, **29**, 1149–1157

Chugani, H. T., Phelps, M. E. and Mazziotta, J. C. (1987), 'Positron emission tomography study of human brain functional development', *Annals of Neurology*, **22**, 487–497

Conel, J. L. (1939-67), *The Postnatal Development of the Human Cerebral Cortex*, Vols 1–8, Cambridge: Harvard University Press

Dobbing, J. (1974), 'Later development of the brain and its vulnerability', in *Scientific Basis of Paediatrics*, eds Davis, J. A. and Dobbing, J., London: Heinemann

Howard, E., Granoff, D. and Bujnovsky, P. (1969), 'DNA, RNA and cholesterol increases in cerebrum and cerebellum during development of human fetuses', *Brain Research*, **14**, 697–706

Hubel, D. H. and Wiesel, T. N. (1963), 'Receptive fields of cells in striate cortex of very young visually inexperienced kittens', *Journal of Neurophysiology*, **26**, 996–1002

Huttenlocher, P. R. and Courten, Ch. de (1987), 'The development of synapses in striate cortex of man', *Human Neurobiology*, **6**, 1–10

Koff, E., Naeser, M. A., Pieniadz, J. M., Foundas, A. L. and Levine, H. L. (1986), 'Computed tomographic scan hemispheric asymmetries in right and left handed male and female subjects', *Archives of Neurology*, **43**, 487–493

Koop, M., Rilling, G., Herrman, A. and Kretschmann, H.-J. (1986), 'Volumetric development of the fetal telencephalon, cerebral cortex, diencephalon and rhombencephalon including the cerebellum in Man', *Bibliotheca Anatomica*, **28**, 53–78

Kretschmann, H.-J., Kammradt, G., Kranthausen, I., Sauer, B. and Wingert, F. (1986a), 'Brain growth in Man', *Bibliotheca Anatomica*, **28**, 1–26

Kretschmann, H.-J., Kammradt, G., Kranthausen, I., Sauer, B. and Wingert, F. (1986b), 'Growth of the hippocampal formation in Man', *Bibliotheca Anatomica*, **28**, 27–52

Kurz, E. M., Sengelaub, D. R. and Arnold, A. P. (1986), 'Androgens regulate the dentritic length of mammalian motoneurons in adulthood', *Science*, **232**, 395–397

Levy, J. (1974), 'Psychobiological implications of bilateral asymmetry', in *Hemisphere Functions in the Human Brain*, eds Dimond, S. J. and Beaumont, J., pp. 121–183, London: Elek

Mann, M. D. (1984), 'The growth of the brain and skull in children', *Developmental Brain Research*, **13**, 169–178

Merrill, E. G. and Wall, P. D. (1972), 'Factors forming the edge of a receptive field; the presence of relatively ineffective afferent terminals', *Journal of Physiology*, **226**, 825–846

Meire, H. B. (1986), 'Ultrasound measurement of fetal growth', in *Human Growth*, 2nd edn, Vol 1, eds Falkner, F. and Tanner, J. M., New York: Plenum

Nottebohm, F. (1981), 'A brain for all seasons: cyclical anatomical changes in song control nuclei of the canary brain', *Science*, **214**, 1368–1369

Patel, A. J., Balazs, R., Altman, J. and Anderson, W. J. (1975), 'Effect of X-irradiation on the biochemical maturation of rat cerebellum: postnatal cell formation', *Radiation Research*, **62**, 470–477

Purpura, D. P. (1975), 'Morphogenesis of visual cortex in the preterm infant', in *Growth and Development of the Brain*, ed. Brazier, M. A. B., New York: Raven

Pysh, J. J. and Weiss, G. M. (1979), 'Excercise during development induces an increase in Purkinje cell dendritic tree size', *Science*, **206**, 230–231

Rabinowicz, T. (1986), 'The differentiate maturation of the human cerebral cortex', in *Human Growth*, 2nd edn, Vol 2, eds Falkner F. and Tanner, J. M., New York: Plenum

Rakic, P. (1976), 'Prenatal genesis of connections subserving ocular dominance in the rhesus monkey', *Nature*, **261**, 467–471

Rakic, P. and Riley, K. P. (1983), 'Overproduction and elmination of retinal axons in the fetal rhesus', *Science*, **219**, 1441–1443

Schultz, D. M., Gardiano, D. A. and Schultz, D. H. (1962), 'Weights of organs of fetuses and infants', *Archives of Pathology*, **74**, 244–250

Simmons, D. J. and Lund, P. W. (1981), 'Early experience of tactile stimulation influences organization of somatic sensory cortex', *Nature*, **326**, 694–697

Slatkin, D. N., Friedman, L., Irsa, A. P. and Micca, P. A. (1985), 'The stability of DNA in human cerebellar neurons', *Science*, **228**, 1002–1005

Tanner, J. M. (1977), *Education and Physical Growth*, 2nd edn, London: Hodder and Stoughton

Wiesel, T. N. and Hubel, D. H. (1963), 'Effects of visual deprivation on morphology and physiology of cells in the cat's lateral geniculate body', *Journal of Neurophysiology*, **26**, 978–993

Witelson, S. F. (1976), 'Sex and the single hemisphere: specialization of the right hemisphere for spatial processing', *Science*, **193**, 425–427

Witelson, S. F. (1977), 'Developmental dyslexia: two right hemispheres and none left', *Science*, **195**, 309–311

Yakovlev, P. I. and Lecours, A. R. (1967), 'The myelogenetic cycles of regional maturation of the brain', in *Regional Development of the Brain in Early Life*, ed. Minkowski, A., Oxford: Blackwell Scientific Publications

Chapter 9

Further Reading

Bielicki, T. (1986), 'Physical growth as a measure of the economic well-being of populations; the twentieth century', in *Human Growth*, 2nd edn, Vol 3 eds Falkner, F. and Tanner, J. M., New York: Plenum

An excellent introduction to the use of growth data as an instrument for monitoring the economic and social well-being of modern populations. Includes description of the particularly comprehensive Polish studies of social class and height and age at menarche

Brundtland, G. H., Liestøl, K. and Walløe, L. (1980), 'Height, weight and menarcheal age of Oslo schoolchildren during the last 60 years', *Annals of Human Biology*, **7**, 307–322

A comprehensive and elegant paper on the secular trend and on class differences in Norway, which boasts the most complete records in Europe over the last 60 years

Elwood, P. C., Sweetnam, P. M., Gray, O. P., Davies, D. P. and Wood, P. D. P. (1987), 'Growth of children from 0-5 years with special reference to mother's smoking in pregnancy', *Annals of Human Biology*, **14**, 543–558

A particularly well planned study, of 1,163 consecutive pregnancies in South Wales. Has important data on the age at which social class differences develop, as well as on smoking

Eveleth, P. B. and Tanner, J. M. (1990), *Worldwide Variation in Human Growth*, 2nd edn, Cambridge: University Press

A massive compilation of all known world data on children's growth originally made as part of the International Biological Programme in 1964–74 but in the second edition covering 1974–1988. Though essentially a reference work, the text and figures deal quite succinctly with the environmental and genetic sources of differences between growth in different populations. Much of the documentation for the statements in this chapter is to be found here

Fogel, R. W. (1986), 'Physical growth as a measure of the economic well-being of populations: the eighteenth and nineteenth centuries', in *Human Growth*, 2nd edn, Vol 3, eds Falkner, F. and Tanner, J. M., New York: Plenum

A description of the use of growth data to elucidate aspects of the economic history of the nineteenth century. Discusses data on American slaves, Civil War soldiers and National Guardsmen, going back to 1710

Garn, S. M. and Bailey, S. M. (1986), 'The genetics of maturation', in *Human Growth*, 2nd edn, Vol 3, eds Falkner, F. and Tanner, J. M., New York: Plenum

A review of the genetic-environment interaction in respect of the timing of events such as the appearance of ossification centres, the eruption of teeth, and the events of sexual maturation

Garn, S. M. and Clark, D. C. (1975), 'Nutrition, growth, development and maturation: findings from the ten-state nutrition survey of 1968–1970', *Pediatrics*, **56**, 309–316

A description of the most comprehensive attempt yet made to survey the nutritional status of the population of the United States. Documents exceptionally well the relation between income and growth rate

Goldstein, H. (1971), 'Factors influencing the height of seven-year-old children: results from the National Child Development Study', *Human Biology*, **43**, 92–111

A thorough analysis of the variation in heights of children in a truly national sample (all births in one week of 1958). Suitable only for readers whose statistical knowledge embraces elementary analysis of variance technique

Goldstein, H. and Tanner, J. M. (1980), 'Ecological considerations in the creation and use of growth standards', *Lancet*, **i**, 582–585

A discussion of the issues involved in growth monitoring of populations especially the question of local versus universal standards

Greene, L. S. and Johnston, F. E., eds (1980), *Social and Biological Predictors of Nutritional Status, Physical Growth and Neurological Development*, New York: Academic Press

The 13 chapters in this book, contributed by 11 authors, constitute the best collection yet assembled on this topic. A very useful introduction to the original literature and the views of outstanding authorities

Greulich, W. W. (1976), 'Some secular changes in the growth of American-born and native Japanese children', *American Journal of Physical Anthropology*, **45**, 553–568

A comparison not only of San Francisco-born and -dwelling with Japan-born and -dwelling Japanese, but of the relative secular trends in the two groups, showing how the Japan-born have caught up the California-born, in whom the trend stopped in the 1950s

Hiernaux, J. (1968), 'Bodily shape differentiation of ethnic groups and of the sexes through growth', *Human Biology*, **40**, 44–62

An important paper both as to method and result. Shows how to follow the development of shape during growth, and how misleading indices such as sitting height/stature may be

Lindgren, G. (1976), 'Height, weight and menarche in Swedish urban school children in relation to socio-economic and regional factors', *Annals of Human Biology*, **3**, 510–528

The first paper to report absence of effects of social class on height growth. A landmark in growth as a measure of the classlessness of society

Meredith, H. V. (1976), 'Findings from Asia, Australia, Europe and North America on secular change in mean height of children, youths and young adults', *American Journal of Physical Anthropology*, **44**, 315–326

A review of the secular change in height, paying particular attention to the data on conscripted soldiers and stressing the part played by earlier completion of height

Mueller, W. H. (1986), 'The genetics of size and shape in children and adults', in *Human Growth*, 2nd edn, Vol 3, eds Falkner, F. and Tanner, J. M., New York: Plenum

An up-to-date review of the genetics of size and shape, particularly emphasising the environment–genetic interaction during the growing period

Richardson, S. A. (1975), 'Physical growth of Jamaican school children who were severely malnourished before two years of age', *Journal of Biosocial Science*, **7**, 445–462

Besides reporting a most careful study, this paper reviews critically the results of previous malnutrition studies and discusses research designs and the generalizations which may currently be made. For the most part non-technical and highly recommended

Schreider, E. (1964), 'Recherches sur la stratification sociale des caracteres biologiques', *Biotypology*, **25**, 105–135

A comprehensive review of relations between height, body build, mental ability and social class, in the UK and France

Stein, Z., Susser, M., Saenger, G. and Marolla, F. (1975), *Famine and Human Development: The Dutch Hunger Winter of 1944–1945*, New York: Oxford University Press

A very important (and very readable) account of the follow-up as adults of persons who experienced a brief period of severe famine as foetuses or infants and were subsequently fed at the usual Western-industrial level. The results emphasize the great resilience of the children who survived

Tanner, J. M. (1966), 'Galtonian eugenics and the study of growth. The relation of body size, intelligence test score and social circumstances in children and adults', *Eugenics Review*, **58**, 122–135; reprinted in: *Trends and Issues in Developmental Psychology*, eds Mussen, P. H., Largen, J. and Covington, M., New York: Holt, 1969

A review of data from the early studies in the 1890s down to 1966. Discusses in particular the effect of social mobility on distribution of height

Tanner, J. M. (1981), *A History of the Study of Human Growth,* Cambridge: University Press

Discusses in detail the older growth data and the basis and extent of the secular trend

Tanner, J. M. (1986a), 'Growth as a mirror of the condition of society: secular trends and class distinctions', in *Human Growth: A Multi-Disciplinary Review,* ed. Demirjian, A., pp. 3–34, London: Taylor and Francis

An introduction to the use of growth data to investigate the economic, social and political structure of populations, now and in the past

Tanner, J. M. (1986b), 'Physical development', in *Childhood Epidemiology,* eds Alberman, E. D. and Peckham, C. A., *British Medical Bulletin,* **42,** 131–138

A discussion of the use of auxological data for monitoring the health of population groups, contrasted with monitoring the growth of individuals

Thoday, J. M. (1965), 'Geneticism and environmentalism', in *Biological Aspects of Social Problems,* eds Meade, J. E. and Parkes, A. S., London: Oliver and Boyd

A sharp and accurate statement of the interaction of genetic and environmental influences in the moulding of phenotypic characteristics. Full of wisdom; the best introduction for the non-biologist to the nature-nurture problem

Waterlow, J. C. and Payne, P. R. (1975), 'The protein gap', *Nature,* **258,** 113–117

A most salutary review of the present situation in the world regarding infant malnutrition. The authors discount the notion that the lack is of protein and contend it is of total food, or energy. A good introduction to nutritional energetics

Waterlow, J. C. ed.(1988) *Linear growth retardation in less developed countries,* New York: Raven

A collection of papers given at a symposium of epidemiologists, nutritionists, paediatricians and auxologists. Particularly useful papers on the mathematics of screening and early diagnosis by Healy et al., on poverty by Martorell et al., and on infection by Nabarro et al., and Tomkins

van Wieringen, J. C. (1978), 'Secular growth changes', in *Human Growth,* 2nd edn, Vol 3, eds Falkner F. and Tanner, J. M., New York: Plenum

A detailed review of the subject, with representation of the very full Dutch records for height of conscripts during the last hundred years. Emphasizes the detailed relationships between environmental factors and changes in heights and weights

Wilson, R. S. (1979), 'Twin growth: initial deficit, recovery, and trends in concordance from birth to nine years', *Annals of Human Biology,* **6,** 205–220

A report which details the growth of the Louisville twins up to age nine. Thorough and very readable

References

Babson, S. G. and Phillips, D. S. (1973), 'Growth and development of twins dissimilar in size at birth', *New England Journal of Medicine*, **289**, 937–40

Brudevoll, J. E., Liestøl, K. and Walløe, L. (1979), 'Menarcheal age in Oslo during the last 140 years', *Annals of Human Biology*, **6**, 407–416

Bielicki, T. and Charzewski, J. (1983), 'Body height and upward social mobility', *Annals of Human Biology*, **10**, 403–408

Billewicz, W. Z. and McGregor, I. A. (1982), 'A birth-to-maturity longitudinal study of heights and weights in two West African (Gambian) villages, 1951–1975', *Annals of Human Biology*, **9**, 309–320

Billewicz, W. Z., Fellowes, H. M. and Thomson, A. M. (1981), 'Menarche in Newcastle-upon-Tyne girls', *Annals of Human Biology*, **8**, 313–320

Billewicz, W. Z., Thomson, A. M. and Fellowes, H. M. (1983), 'A longitudinal study of growth in Newcastle-upon-Tyne adolescents', *Annals of Human Biology*, **10**, 125–134

Cheek, D. B., Holt, A. B., London, W. T., Ellenberg, J. H., Hill, D. E. and Sever, J. L. (1976), 'Nutritional studies in the pregnant rhesus monkey: the effect of protein-calorie or protein deprivation on growth of the fetal brain', *American Journal of Clinical Nutrition*, **29**, 1149–1157

Chinn, S. and Rona, R. J. (1984), 'The secular trend in the height of primary school children in England and Scotland from 1972 to 1980', *Annals of Human Biology*, **11**, 1–16

Cliquet, R. L. (1968), 'Social mobility and the anthropological structure of populations', *Human Biology*, **40**, 17–43

Dahlman, N. and Petersen K. (1977), 'Influences of environmental conditions during infancy on final body size', *Pediatric Research*, **11**, 695–700

Damon, A. (1968), 'Secular trend in height and weight within Old American families at Harvard 1870–1965. I. Within twelve four-generation families', *American Journal of Physical Anthropology*, **29**, 45–50

Davies, D. P. and Yamamuro, T. (1985), 'Growth of Hong Kong infants during the first two years of life', *Early Human Development*, **11**, 247–254

Duggan, M. B., Alway, J. and Milner, R. D. G. (1986), 'The nutritional cost of measles in Africa', *Archives of Disease in Childhood*, **61**, 61–66

Elias, M. F. and Samonds, K, W. (1977), 'Protein and calorie malnutrition in infant cebus monkeys. Growth and behavioral development during deprivation and rehabilitation', *American Journal of Clinical Nutrition*, **30**, 355–366

Ellison, P. T. (1981), 'Morbidity, mortality and menarche', *Human Biology*, **53**, 635–644

Eveleth, P. B. and Tanner J. M. (1976), *'World-wide variation in human growth'*, 1st edn, Cambridge: University Press

Fischbein, S. (1977a), 'Intra-pair similarity in physical growth of monozygotic and of dizygotic twins during puberty', *Annals of Human Biology*, **4**, 417–430

Fischbein, S. (1977b), 'Onset of puberty in MZ and DZ twins', *Acta Genetica Medica Gemellologica*, **26**, 151–158

Fleagle, J. G., Samonds, K. W. and Hegsted, D. M. (1975), 'Physical growth of cebus monkeys, *Cebus albifrons*, during protein or calorie deficiency', *American Journal of Clinical Nutrition*, **28**, 246–253

Friend, G. E. and Bransby, E. R. (1947), 'Physique and growth of schoolboys', *Lancet*, **2**, 677

Goldstein, H. and Peckham, C. (1976), 'Birthweight, gestation, neonatal mortality and child development', in *The Biology of Fetal Growth*, eds Roberts, D. F. and Thomson, A. M., *Symposia of the Society for the Study of Human Biology*, **15**, 81–108

Greksa, L. P. (1988), 'Effect of altitude on the stature, chest depth and forced vital capacity of low-to-high altitude migrant children of European ancestry', *Human Biology*, **60**, 23–32

Hamill, P. V., Johnston, F. E. and Lemeshow, S. (1972), 'Height and Weight of Children: Socio-Economic Status: United States', U.S. Dept. of Health, Education and Welfare, Publn No. HRA 73–1601, Rockville, Md.

Hamill, P. V., Johnston, F. E. and Lemeshow, S. (1973), 'Body Weight, Stature and Sitting Height: White and Negro Youths 12–17 Years', U.S. Dept. of Health, Education and Welfare, Publn No. HRA–74–1608, Rockville, Md.

Hermanussen, M., Geiger-Benoit, K. and Sippell, W. G. (1986), 'GH deficient children receiving GH replacement do not grow during intermittent infectious illness', *Acta Paediatrica Scandinavica*, **75**, 601–604

Hill, E. E., Mayers, R. E., Holt, A. B., Scott, R. E. and Cheek, D. B. (1971), 'Fetal growth retardation produced by experimental placental insufficiency in the rhesus monkey. 2. Chemical composition of the brain, liver, muscle and carcass', *Biology of the Neonate*, **19**, 68–82

Janes, M. D. (1975), 'Physical and psychological growth and development', *Environmental Child Health*, **121**, 26–30

Johnston, F. E., Hamill, P. V. V. and Lemeshow, S. (1974), 'Skinfold thickness in a national probability sample of U.S. males and females aged 6 through 17 years', *American Journal of Physical Anthropology*, **40**, 321–324

Jones, D. L., Hemphill, W. and Meyers, E. S. A (1973), 'Height, Weight and Other Physical Characteristics of New South Wales Children. I. Children aged 3 Years and Over'. Sydney, New South Wales: Department of Health

Jones, D. Y., Nesheim, M. C. and Habicht, J. -P. (1985), 'Influences on child growth associated with poverty in the 1970's: an examination of HANES I and HANES II. cross-sectional US national surveys', *American Journal of Clinical Nutrition*, **42**, 714–724

Kerr, G. R., el Lozy, M. and Scheffler, G. (1975), 'Malnutrition studies in *Macaca mulatta*. IV Energy and protein consumption during growth failure and "catch-up" growth', *American Journal of Clinical Nutrition*, **28**, 1364–1376

Kiil, V. (1939), 'Stature and growth of Norwegian men during the last 200 years', *Skrifte Norske Videnskap Akademie*, No.6, 175 pp.

Kimura, K. (1983), 'Skeletal maturity and bone growth in twins', *American Journal of Physical Anthropology*, **60**, 491–497

Laska-Mierzejewska, T., Milicer, H. and Piechaczek, H. (1982), 'Age at menarche and its secular trend in urban and rural girls in Poland', *Annals of Human Biology*, **9**, 227–234

Lamb, H. H. (1973), 'Whither climate now?' *Nature*, **244**, 395–397

Lamb, W. H., Foord, F., Lamb, C. and Whitehead, R. G. (1984), 'Changes in maternal and child mortality rates in three isolated Gambian villages over 10 years', *Lancet*, **2**, 912–914

Lee, P. A. (1980), 'Independence of seasonal variation of growth from temperature change', *Growth,* **44**, 54–59

Lerberghe, W. van (1988), 'Linear growth retardation and mortality', in *Linear Growth Retardation in Less Developed Countries,* pp. 245–264, ed. Waterlow, J. C., New York: Raven

Ljung, B. O., Bergsten-Brucefors, A. and Lindgren, G. (1974), 'The secular trend in physical growth in Sweden', *Annals of Human Biology,* **1**, 245–256

Low, W. D., Kung, L. S. and Leong, J. C. Y. (1982), 'Secular trend in the sexual maturation of Chinese girls', *Human Biology,* **54**, 539–550

MacMahon, B. (1973), 'Age at Menarche, United States', U.S. Dept of Health, Education and Welfare, Publn No. HRA 74–1615, Rockville, Md.

Malcolm, L. A. (1971), 'Growth and Development in New Guinea', Madang: Institute of Human Biology (monograph No. 1)

Marshall, W. A. (1971), 'Evaluation of growth rate in height over periods of less than a year', *Archives of Disease in Childhood,* **46**, 414–420

Marshall, W. A. and Swan, A. V. (1971), ' Seasonal variation in growth rates of normal and blind children', *Human Biology,* **43**, 502–516

Martorell, R., Yarbrough, C., Yarbrough, S. and Klein, R. E. (1980), 'The impact of ordinary illnesses on the dietary intakes of malnourished children', *American Journal of Clinical Nutrition,* **33**, 345–350

Martorell, R., Yarbrough, C. and Klein, R. E. (1980), 'Protein-calorie supplementation, body size, and skeletal maturity in three-year-old Guatamalan children', *Studies in Physical Anthropology,* **6**, 23–32

Martorell, R., Mendoza, F. S., Castillo, R. O., Pawson, I. and Gand-Budge, C. C. (1987), 'Short and plump physique of Mexican-American children', *American Journal of Physical Anthropology,* **73**, 475–488

Martorell, R., Malina, R. M., Castillo, R. O., Mendoza, F. S. and Pawson, I. G. (1988), 'Body proportions in three ethnic groups: children and youths 2–17 years in NHANES II and HHANES', *Human Biology,* **60**, 205–222

McManus, I. C. and Mascie-Taylor, C. G. N. (1984), 'Human assortative mating for height: non-linearity and heteroscedasticity', *Human Biology,* **56**, 617–624

Mora, J. O., Herrera, M. G., Suescun, J., de Navarro, L. and Wagner, M. (1981), 'The effects of nutritional supplementation on physical growth of children at risk of malnutrition', *American Journal of Clinical Nutrition,* **34**, 1885–1892

Nagai, N., Matsumoto, K., Mino, T., Takeuchi, H. and Takeda, S. (1980), 'The secular trends in the menarcheal age and the maximum growth age in height for Japanese schoolgirls', *Wakayama Medical Reports,* **23**, 41–45

Panek, S. and Piesecki, E. (1971), 'Nowa Huta: integration of the population in the light of anthropological data', *Materialy i Prace Antropologiczne,* **80**, 1–249

Peckham, C., Butler, N. and Frew, R. (1983), 'Medical and social aspects of children with educational problems', in *Growing up in Great Britain,* pp. 181-188, ed. Fogelman, K., London: MacMillan

Prokopec, M. (1985), 'Validity period of national growth standards', *Acta Medica Auxologica,* **17**, 125–140

Reves, R. (1985), 'Declining fertility in England and Wales as a major cause of the 20th century decline in mortality', *American Journal of Epidemiology,* **122**, 112–126

Richardson, S. A., Koller, H., Katz, M. and Albert, K. (1978), 'The contributions of differing degrees of acute and chronic malnutrition to the intellectual development of Jamaican boys', *Early Human Development*, **2**, 163–170

Roberts, D. F., Danskin, M. J. and Chinn S. (1975), 'Menarcheal age in Northumberland', *Acta Paediatrica Scandinavica*, **64**, 845–852

Roede, M. J. and van Wieringen, J. C. (1985), 'Growth Diagrams 1980. Netherlands Third Nationwide Survey', *Tijdschrift voor Sociale Gezondheitszong*, **63**, suppl. 34 pp.

Roland, M. G. M., Cole, T. J. and Whitehead, R. G. (1977), 'A quantitative study into the role of infection in determining nutritional status in Gambian village children', *British Journal of Nutrition*, **37**, 441–450

Roland, M. G. M., Roland, S. G. J G., and Cole, T. J. (1988), 'Impact of infection on the growth of children from 0 to 2 years in an urban West African Community', *American Journal of Clinical Nutrition*, **47**, 134–138

Rona, R. J. and Chinn, S. (1984), 'The National Study of Health and Growth: nutritional surveilance of primary school children from 1972 to 1981 with special reference to unemployment and social class', *Annals of Human Biology*, **11**, 17–28

Rona, R. J. and Chinn, S. (1986), 'National Study of Health and Growth: social and biological factors associated with height of children from ethnic groups living in England', *Annals of Human Biology*, **13**, 453–472

Rona, R. J. and Florey, C. du V. (1980), 'National Study of Health and Growth: respiratory systems and height in primary schoolchildren', *International Journal of Epidemiology*, **9**, 35–43

Rona, R. J., Swan, A. V. and Altman, D. G. (1978), 'Social factors and height of primary schoolchildren in England and Wales', *Journal of Epidemiology and Community Health*, **32**, 147–154

Rosenbaum, S., Skinner, R. K. Knight, I. B. and Garrow, J. W. (1985), 'A survey of heights and weights of adults in Great Britain, 1980', *Annals of Human Biology*, **12**, 115–127

Rosetta, L. (1988), 'Seasonal changes and the physical development of young Serere children in Senegal', *Annals of Human Biology*, **15**, 179–190

Rutter, M., Tizard, J. and Whitmore, K. eds (1970), *Education, Health and Behaviour*, London: Longman

Shields, J. (1962), *Monozygotic Twins Brought Up Apart and Brought Up Together*, London: Oxford University Press

Smith, A. M., Chinn, S. and Rona, R. J. (1980), 'Social factors and height gain of primary schoolchildren in England and Scotland', *Annals of Human Biology*, **7**, 115–124

Smith, D. W., Truog, W., Rogers, J. E., Greitzler, L. J., Skinner, L., McCann, J. J. and Harvey, M. A. S. (1976), 'Shifting linear growth during infancy: illustration of genetic factors in growth from fetal life through infancy', *Journal of Pediatrics*, **89**, 225–230

Stini, W. A. (1972), 'Malnutrition, body size and proportion', *Ecology of Food and Nutrition*, **1**, 121

Stini, W. (1986), 'Reductions in body size and the preservation of genetic variability in tropical populations', in *Genetic Variation and its Maintenance*, eds Roberts, D. F. and de Stefano, G. F., Cambridge: University Press, *Symposia of the Society for the Study of Human Biology,* **27**, 253–262

Stini, W., Weber, C. W., Kemberling, S. R. and Vaughan, L. A. (1980), 'Lean tissue growth and disease susceptibility in bottle-fed versus breast-fed infants', in *Social and Biological Predictors of Nutritional Status, Physical Growth, and Neurological Development,* pp. 61–80, eds Green, L. S. and Johnston, F. E., New York: Academic Press

Sundelin, C. and Vuille, J.-C. (1979), 'Socio-economic status of the family and health problems in the preschool child. Lack of correlation in a Swedish sample', *Acta Paediatrica Scandinavica,* Supplement, **275**, 59–65

Susanne, C. (1975), 'Genetic and environmental influence on morphological characteristics', *Annals of Human Biology,* **2**, 279–288

Tanner, J. M. (1962), *Growth At Adolescence,* 2nd edn, Oxford: Blackwell Scientific Publications

Tanner, J. M. (1964), *The Physique of the Olympic Athlete,* London: Allen & Unwin; distributed by Institute of Child Health, 30 Guilford Street, London WC1

Thomson, A. M. (1959), 'Maternal stature and reproductive efficiency', *Eugenics Review,* **51**, 157–162

Valman, H. B. (1974), 'Intelligence after malnutrition caused by neonatal resection of ileum', *Lancet,* **1**, 425–427

Waaler, H. Th. (1984), 'Height, weight and mortality: the Norwegian experience', *Acta Medica Scandinavica,* Supplement, **679**, 1–56

Wenlock, R. W., Disselduff, M. M., Skinner, R. K. and Knight, I. (1986), *The Diets of British Schoolchildren,* London: HMSO

Whitehead, R. G. and Paul, A. A. (1984), 'Growth charts and the assessment of infant feeding practices in the Western world and in developing countries', *Early Human Development,* **9**, 187–208

Wilson, R. S. (1976), 'Concordance in physical growth for monozygotic and dizygotic twins', *Annals of Human Biology,* **3**, 1–10

Winick, M., Rosso, P. and Waterlow, J. (1970), 'Cellular growth of cerebrum, cerebellum and brain stem in normal and marasmic children', *Experimental Neurology,* **26**, 393

Wyshak, G. (1983), 'Secular changes in age at menarche in a sample of U.S. women', *Annals of Human Biology,* **10**, 75–78

Chapter 10

Further Reading

Hubel, D. H. and Wiesel, T. N. (1970), 'The period of susceptibility to the physiological effects of unilateral eye closure in kittens', *Journal of Physiology,* **206**, 419–436

Describes very clearly waxing and waning of the susceptibility between 4 and 12 weeks after the kitten's birth, and relates this to the development of cells in the brain

Prader, A., Tanner, J. M. and von Harnack, G. A. (1963), 'Catch-up growth following illness or starvation', *Journal of Pediatrics*, **62**, 646–659

The original description and designation of catch-up growth. Simple, and requires no special medical knowledge

Scott, J. P., Stewert, J. M., De Ghett, V. J. (1974), 'Critical periods in the organization of systems', *Developmental Psychobiology*, **7**, 489–513

A general review of the 'sensitive period' theory, with helpful diagrams. Examples are given from socialization in the puppy

Tanner, J. M. (1963), 'The regulation of human growth', *Child Development*, **34**, 817–847 (see also *Nature*, **199**, 845–850, 'Regulation of growth in size of mammals'.)

Describes a tentative theory of how the regulation of postnatal growth in man works. Suggests that the organism is able to monitor both its actual and its 'target' size and grow in accordance with the mismatch between them

Mosier, H. D. (1989), 'Catch-up growth and target size in experimental animals', in *The Physiology of Growth* pp. 30–45, eds Tanner, J. M. and Preece, M. A., Cambridge: University Press

A summary of the author's experimental work on catch-up growth in rats

References

Harweith, R. S., Smith, E. L. III, Duncan, G. C., Crawford, M, L, J, and von Noorden, G. K. (1986), 'Multiple sensitive periods in the development of the primate visual system', *Science*, **232**, 235–237

Jacobson, M. (1970) *Developmental Neurobiology*, New York: Holt, Rhinehart and Winston

Lorenz, K. (1960), in *Discussions on Child Development*,Vol 4, p. 128, eds Tanner, J. M. and Inhelder, B., London: Tavistock Publications

McGraw, M. (1943), *The Neuromuscular Maturation of the Human Infant*, New York: Columbia University Press

Money, J. and Hampson, J. G. (1955), 'Idiopathic sexual precocity in the male', *Psychosomatic Medicine*, **17**, 1–15

Mosier, H. D. Jr., Jansons, R. A., Swingle, K. F., Sondhaus, C. A., Dearden, L. C. and Halsall, L. C. (1985), 'Growth hormone secretion in the stunted head-irradiated rat', *Pediatric Research*, **19**, 543–548

Mosier, H. D. Jr., Jansons, R. A., Swingle, K. F. and Dearden, L. C. (1986), 'Dissociation of catch-up growth control and neural control of growth hormone secretion in the stunted head-irradiated rat', *Pediatric Research*, **20**, 261–264

Tanner, J. M. (1975), 'Towards complete success in the treatment of growth hormone deficiency; a plea for earlier ascertainment', *Health Trends*, **7**, 61 65

Waddington, C. H. (1957), *The Strategy of the Genes*, London: Allen and Unwin

Chapter 11

Further Reading

Bayley, N. and Pinneau, S. R. (1952), 'Tables for predicting adult height from skeletal age: revised for use with the Greulich–Pyle hand standards', *Journal of Pediatrics*, **40**, 423–441 and **41**, 371

Describes the system of height prediction based on the Greulich–Pyle method for bone age

Buckler, J. M. H. (1979), *A Reference Manual of Growth and Development*, Oxford: Blackwell Scientific Publications

A manual which contains a useful collection of charts from various sources

Hamill, P. V. V., Drizd, T. A., Johnson, C. L., Reed, R. R. and Roche, A. F. (1976), 'NCHS growth charts, 1976', *Vital Statistics Report*, HRA 76–1120, Vol 25, No. 3, supplement, Rockville, Md.

A general description of the U.S.A. cross-sectional growth charts, including height and weight for age; and weight for height, prepuberty ages pooled

Healy, M. J. R. (1974), 'Notes on the statistics of growth standards', *Annals of Human Biology*, **1**, 41–46

A very simply written account of the use and meaning of standards for growth; the best introduction to the subject

Tanner, J. M. (1986), 'Use and abuse of growth standards', in *Human Growth*, 2nd edn, Vol 3, eds Falkner, F. and Tanner, J. M., New York: Plenum

A description of the different sets of standards, cross-sectional, longitudinal and conditional with consideration of the use to which each may be put

Tanner, J. M. and Davies, P. S. W. (1985), 'Clinical longitudinal standards for height and height velocity for North American children', *Journal of Pediatrics*, **107**, 317–329

Describes the way in which the North American longitudinal standards were constructed, using the shape of the individual curve, and the amplitude derived from the cross-sectional studies of the NCHS. Simpler to follow than the Tanner et al., (1966) paper

Tanner, J. M., Whitehouse, R. H. and Takaishi, M. (1966), 'Standards from birth to maturity for height, weight, height velocity, and weight velocity; British children 1965', *Archives of Disease in Childhood*, **41**, 454–471; 613–635

The basic paper on growth standards, in which longitudinal standards based on both longitudinal data (for shape) and cross-sectional data (for amplitude) were introduced. Not light reading, but quite understandable to readers with an elementary knowledge of statistics

Tanner, J. M., Whitehouse, R. H., Cameron, N., Marshall. W. A., Healy, M. J. R. and Goldstein, H. (1983), *Assessment of Skeletal Maturity and Prediction of Adult Height*, 2nd edn, London: Academic Press

Gives illustrations of how bone age is assessed, and tables for predicting adult height.
Before age 6 in girls and 8 in boys the tables do not involve bone age

Waterlow, J. C., Buzina, R., Keller, W., Lane, J. M., Nichaman, M. Z. and Tanner,
J. M. (1978), 'The presentation and use of height and weight data for comparing
the nutritional status of groups of children under the age of 10', *Bulletin of the
World Health Organisation*, **55**, 489–498

*The U.S.A. National Center for Health Statistics charts are presented in this paper, which
also contains a very clear discussion of the role of anthropometric measurements in
nutritional surveillance*

References

Baumgartner, R. N., Roche, A. F. and Himes, J. H. (1986), 'Incremental growth
 tables: supplementary to previously published charts', *American Journal of
 Clinical Nutrition*, **43**, 711–722
Berkey, C. S., Reed, R. B. and Valadian, I. (1983), 'Longitudinal growth standards
 for preschool children', *Annals of Human Biology*, **10**, 57–68
Billewicz. W. Z., Thomson, A. M. and Fellowes, H. M. (1983), 'Weight-for-
 height in adolescence', *Annals of Human Biology*, **10**, 119–124
Brown, K. H., Black, R. E., Becker, S. and Hoque, A. (1982), 'Patterns of physical
 growth in a longitudinal study of young children in rural Bangladesh',
 American Journal of Clinical Nutrition, **36**, 294–302
Buckler, J. M. H. (1978), 'Variations in height throughout the day', *Archives of
 Disease in Childhood*, **53**, 762
Cole, T. J. (1985), 'A critique of the NCHS weight for height standards', *Human
 Biology*, **57**, 183–196
CusChlo, J. S. and Nube, M. (1981) 'Attainable growth', *Lancet*, **2**, 1223
Goldstein, H. and Tanner, J. M. (1980), 'Ecological considerations in the creation
 and use of growth standards', *Lancet*, **I**, 582–585
Hamill, P. V. V., Johnston, F. E. and Lemeshow, S. (1973), *Height and Weight of
 Youths 12–17 Years: United States*, U.S. Dept. of Health, Education and
 Welfare, Publn. No. HSM 73–1606, Rockville, Md.
Himes, J. H., Roche, A. F. and Thissen, D. (1981), ' Parent-specific adjustments
 for assessment of recumbent length and stature', *Monographs in Pediatrics*,
 13, 88 pp., Basel: Karger
Johnston, F. E., Bogin, B., MacVean, R. B. and Newman, B. C. (1984), 'A
 comparison of international standards versus local reference data for the
 triceps and subscapular skinfolds of Guatemalan children and youth', *Human
 Biology*, **56**, 157–171
Mino, T. (1984), 'Standards for height and annual increment in Japanese children',
 Journal of the Wakayama Medical Society, **35**, 427–443
Persson, L. A. (1985), 'Infant feeding and growth – a longitudinal study in three
 Swedish communities', *Annals of Human Biology*, **12**, 41–52
Prokopec, M. (1985), 'Validity period of national growth standards', *Acta Medica
 Auxologica*, **17**, 125–140

Roche, A. F. and Himes, J. H. (1980), 'Incremental growth charts', *American Journal of Clinical Nutrition*, **33**, 2041–2052

Rona, R. J. and Altman, D. G. (1977), 'The National Study of Health and Growth. Standards of attained height, weight and triceps skinfold in English children 5–11 years old', *Annals of Human Biology*, **4**, 501–524

Ryan, A. S. and Martinez, G. A. (1987), 'Physical growth of infants 7 to 13 months of age – results from a National Survey', *American Journal of Physical Anthropology*, **73**, 449–458

Tanner, J. M. (1981), *A History of the Study of Human Growth*, Cambridge: University Press

Tanner, J. M. and Whitehouse, R. H. (1976), 'Clinical longitudinal standards for height, weight, height velocity and weight velocity and the stages of puberty', *Archives of Disease in Childhood*, **51**, 170–179

van Loon, H., Saverys, V., Vuylsteke, J. P., Vlietinck, R. F. and Eeckels, R. (1986), 'Local versus universal growth standards: the effect of using NCHS as a universal reference', *Annals of Human Biology*, **13**, 347–358

Chapter 12

Further Reading

Lacey, K. A. and Parkin, J. M. (1974), 'The normal short child. A community study in Newcastle-upon-Tyne', *Archives of Disease in Childhood*, **49**, 417–424

A very valuable survey of all children born in Newcastle in 1960 and below the 3rd centile for height at age 10, without any known specific disease as a cause. In most the shortness was familial, in some due to delay in growth, and in a few to psychosocial stress

Rallison, M. L. (1986), *Growth Disorders in Infants, Children and Adolescents*, New York: Wiley

An up-to-date, standard textbook on growth disorders for medical students and paediatricians

Smith, D. W. (1977), *Growth and Its Disorders*, Philadelphia: Saunders

An elementary textbook on disorders of growth. Has good reproductions of the standard growth charts. Written for medical students, but not upsetting to the lay parent

Smith, D. W. (1982), *Recognisable Patterns of Congenital Malformation*, 3rd edn, Philadelphia: Saunders

The medical textbook authority on the rare diseases causing short stature allied with various malformations and deformities. Not to be lightly consulted by the lay parent, for the illustrations may be upsetting, but a bible to paediatricians

Tanner, J. M. and Thomson, A. M. (1970), 'Standards for birth-weight at gestation periods from 32 to 42 weeks, allowing for maternal height and weight', *Archives of Disease in Childhood*, **45**, 566–569

Standards for weight at birth, divided according to sex, length of gestation and whether first- or later-born. The data are from Aberdeen, Scotland

Yudkin, P. L., Aboualfa, M., Eyre, J. A., Redman, C. W. G., and Wilkinson, A. R.(1987) 'New birthweight and head circumference centiles for gestational ages 24 to 42 weeks', *Early Human Development*, **15**, 45–52

The best modern UK standards, based on Oxford new borns

References

Brown, P. (1988), 'Human growth hormone therapy and Creuzfeldt–Jakob disease – a drama in three acts', *Pediatrics*, **81**, 85–92

Hindmarsh, P., Smith, P. J., Brook, C. G. D. and Matthews, D. R. (1987), 'The relationship between height velocity and growth hormone secretion in short prepubertal children', *Clinical Endocrinology*, **27**, 581–592

Morabito, F., Aldegheri, R., Tessari, G., Atterrato, A. and Resentini, M. (1985), 'Auxological results of surgical correction of hypometries in congenital osteodysplasia', *Acta Medical Auxologica*, **17**, 279–285

Raben, M. S. (1958), 'Treatment of a pituitary dwarf with human growth hormone', *Journal of Clinical Endocrinology and Metabolism*, **18**, 901–903

Stanhope, R., Adlard, P., Hamill, G., Jones, J., Skuse, D. and Preece, M. A. (1988), Physiological growth hormone secretion during the recovery from psychosocial dwarfism: a case report', *Clinical Endocrinology*, **28**, 335–340

Stanhope, R. and Preece, M. A. (1988), 'Management of constitutional delay of growth and puberty', *Archives of Disease in Childhood*, **63**, 1104–1110

Tanner, J. M., Lejarraga, H. and Cameron, N. (1975), 'The natural history of the Silver-Russell syndrome: a longitudinal study', *Pediatric Research*, **9**, 611–623

Wilson, D. M., Kei, J., Hintz, R. L. and Rosenfeld, R. G. (1988) 'Effects of testosterone therapy for pubertal delay', *American Journal of Diseases of Children*, **142**, 96–99

Author Index

Akutagawa, E., 117
Altman, D. G., 118, 178

Babson, S. G., 123
Baines, M., 9
Balazs, R., 118
Bayley, N., 221
Becker, L., 108
Bielicki, T., 154
Billewicz, W. Z., 131, 152, 163
Blakemore, C., 116
Boas, F., 75, 221
Bock, D., 9
Bransby, E. R., 148
Brasel, J.-A., 30
Breuer, A., 107
Briend, A., 42
Brook, C. G. D., 102
Brown, K. H., 210
Brudevoll, J. E., 160
Brundtland, G.-H., 155–6, 157

Cajal, R., 114
Charzewski, J., 154
Cheek, D., 105, 137
Chinn, S., 159, 164, 199, 207
Chungani, H., 113
Cliquet, R. L., 153
Cole, T. J., 200
Conel, J. L., 108–112

Damon, A., 159
Dargassies, S. St.-A., 50
Davies, P. S. W., 191
Demerjian, A., 79
Descartes, R., 99–100
Deter, R., 47
Dreyfuss-Brisac, C., 50
Drillien, C. M., 49
Dubowitz, V., 50
Duggan, M. B., 133

Elias, M. F., 139
Ellison, P. T., 161
Elwood, P. C., 134, 147, 150–1
Eveleth, P., 140, 144, 156

Fels Research Institute, 11, 78, 202
Fischbein, S., 128
Fleagle, J. G., 137–39

Florey, C. du V., 147
Friend, G. E., 148

Garn, S., 83, 151
Gasser, T., 10
Goeschen, K., 50
Goldstein, H., 151, 211
Greksa, L. P., 145
Greulich, W., 76, 78, 142

Hamill, P. V., 142
Hampson, J. G., 175
Harwerth, R. S., 173
Hay, E., 40
Health Examination Survey, and
    Health and Nutrition Examination
    Survey (USA), 151, 203
Hegsted, D. M., 137–39
Hermanussen, M., 8, 147
Hill, D., 137–40
Ho, K. Y., 91
Hubel, D. H., 116, 119
Huttenlocher, P., 108

Institute of Child Health (UK), 11, 50,
    103, 221
International Children's Centre, 211

Jacobson, M., 172
Janes, M. D., 131–2
Johnston, F. E., 18, 144, 210
Jones, D. L., 145
Jones, D. Y., 151

Kember, N. F., 33
Kemper, H. C. G., 74
Kerr, G. R., 137–40
Kiil, V., 159
Knobil, E., 89
Knobloch, H., 49
Konishi, M., 117
Kurz, E. M., 118

Lamb, H. H., 162
Lamb, W. H., 133
Land, P. W., 117
Laska-Mierzejewska, T., 160
Lee, P. A., 145
Lerberghe, W. van, 163
Levy, J., 114

Limongi, Y. de, 103
Lindgren, G., 154
Loon, H. van, 210
Lorenz, K., 119, 174
Low, W, D,, 144

McGraw, M., 176
McGregor, I. A., 163
McLean, F. H., 46
McManus, I. C., 124
Mark, R., 117
Marshall, W. A., 103, 145–6
Martorell, R., 133, 144
Mascie-Taylor, C. G. N., 124
Merrill, E. G., 117
Milner, R. D. G., 95, 236
Money, J., 175
Montbeillard, P. G. de, 6–7
Moosa, A., 50
Mora, J. O., 134
Morris, R. W., 199, 207
Mosier, D., 170
Mussen, P., 83

Nagai, N., 144
National Center for Health Statistics
    (USA), 151, 202
National Child Development Survey
    (UK), 150, 151, 156
National Study of Health and Growth,
    151, 159
Nilsen, S. T., 45
Nottebohm, F., 118

O'Malley, B. W., 85

Panek, S., 148–9
Pasamanick, B., 49
Patel, A. J., 118
Paul, A. A., 140
Peckham, C., 156
Phillips, D. S., 123
Piaget, J., 113
Piasecki, E., 148–9
Poissonet, C. M., 40
Preece, M. A., 9
Prokopec, M., 210
Purpura, D. P., 111
Pyle, S. I., 76, 78
Pysh, J. J., 115

Raben, M. S., 232
Rabinowicz, T., 109
Rakic, P., 104, 116

Reves, R., 151
Richardson, S., 135
Roche, A., 78
Roede, M. J., 210
Rona, R. J., 147, 152, 159, 164, 178
Rosenbaum, S., 150
Rosetta, L., 147
Roland, M. G. M., 133

Samonds, K. W., 139
Schrader, W. T., 85
Schultz, D. M., 105
Shields, J., 121–2
Shuttleworth, F., 221
Sidman, R. L., 104
Simons, D. J., 117
Sissons, H. A., 33
Smith, D. W., 37, 126–7
Southgate, D. A. T., 40
Stanhope, R., 102
Stein, Z., 135–6
Stini, W. A., 140
Sundelin, C., 155
Susanne, C., 124–6
Susser, M., 135–6

Tanner, J. M., 18, 76–8, 103, 140, 144,
    148, 156, 190–1, 211, 219
Thomson, A. M., 159

Usher, R. H., 46

Valk, I. M., 8
Vuille, J.-C., 155

Waaler, H. Th., 163
Waddington, C. H., 165
Wales, J. K. H., 95, 236
Wall, P. D., 117
Waterlow, J., 131, 164, 203
Wenlock. R. W., 134, 151, 153
Whitehead, R. G., 140
Whitehouse, R. H., 7, 8, 18, 76–8,
    103, 148, 183, 197, 211
Widdowson, E., 148
Wieringen, J. C. van, 159
Wiesel, T.-N., 116, 119
Wilson, R. S., 123
Winick, M., 30
Witelson, S. F., 114
Wyshak, G., 160

Yakovlev, P. I., 112

Subject Index

Where subjects appear *passim* references are given for definitive and selected entries only. Figures in **bold** type indicate definitions. gr. = growth

abortion: social, 36
spontaneous, 37, 46
achondroplasia, **233**
ACTH, *see under* hormones, specific
adenine, 27
adipose tissue, structure of, 30
adolescence, **58**, 176–7
adrenal cortex, 93
foetal 96–7
adrenal gland, 85, 93
disease of, 229
adrenarche, 96
adrenocorticotrophic hormone, *see*
ACTH *under* hormones, specific
age, bone, *see* bone age
dental: as measure of maturity, 78
correlation with skeletal age, 79
developmental, 75–83
gestational, postfertilization, postmenstrual, **37–43**
alcohol and small-for-dates, 46, 228
aldosterone, 93
amino acids, 28, 84, 115
androgenic hormones, *see under*
hormones
androgyny, **68**
arm, gr. of bones of, 35
arthritis, 95
Assoc. for Research into Restricted
Growth (ARRG), 222, 233, 235
asthma, 95, 236
athletics, effects on muscles, 31
'atlas' bone age assessment, 76
average, moving, 10
axon, **106**–9, 114, 115, 117

baldness, 126
basal energy expenditure, **26**
beard, evolutionary function of, 70, 98
birdsong and development of nervous
system, 117–18
birth as a happening, 48–50
blastocyst, **36**
blood group, inheritance of, 28
bone: age, **76–8**

correlation with beginning of
puberty, 80; dental age, 79;
menarche, 79
and prediction of adult height,
219–21
racial differences, and, 144
standards, 218
disorders, 233–5
epiphyses, 32
gr. of, 31–5, 97
regenerative powers of, 24
structure of, 32
bone-scoring, *see* Tanner–Whitehouse
bone-scoring
brain, 104–18
functional areas, development of,
108–13
gr. curve, 16–17, 104, 105, 106
'injury' at birth, 48–9
maturation and evolution, 23
maturity and initiation of puberty,
101
prenatal development of, and
thyroid hormone, 93
protein turnover in, 115
right and left sides of, 113–14
sensitive periods in development of,
173
tumours and short stature, 229–31
visual system, and, 115–17
wave (encephalogram – EEG),
49–50
breast: areola of, 62, 63
bud, **63**, 64
development
in boys, 62
in girls, 63, 98, 99
late, 82–3; menstruation,
correlation with, 103; in
precocious puberty, 96;
standards, 213–15, 217
feeding, 131
nipple of, 63, 64

calcium, 33, 76

273

turnover, 25
canalization and catch-up, 165–71
cartilage, 32–5, 68, 92, 95, 223, 236
catch-up gr., 89, 93, 118, 133, 135,
    165–171, 232, 233
cell: competence, 171–2
    differentiation 24, see also
        histogenesis, morphogenesis
    division, 27, 28–30, 169–70
    germ c., 27
    nucleus of, in muscles, 29–30
    regeneration, 24–25
    structure 26–28
centile (or percentile), 19–21, 126–7,
    167–9
cerebellum, 105, 108, 112, 115, 118
cerebral cortex, 106–13, 115–17
cerebrum, 104, 105, 107, 108
CGT, see under hormones, specific
Child Growth Foundation (UK), 222
cholesterol turnover, 25
chorionic gonadotrophin, 54, 86
chromosomes, 27
    abnormality, incidence in, 37
    sex, determination by, 51–4
climate: body shape, evolutionary and
    direct effects on, 144–5
    menarche, alleged effect on, 145 and
        secular trend, 162
CLIP, see under hormones, specific
clock: biological, and birth, 48–50; and
    puberty, 102–3
    cellular, 30–31, 118
clone, 32
codon, 28
coeliac disease, 235
compensatory growth, 165–6
composition, body, 16–18, 58, 65,
    70–3, 105, 144
computer-aided: auxology, 9
    reconstruction of brain, 104
connective tissue, 30
corpus luteum, 99
corticoids, see glucocorticoids
corticotrophin–releasing factor (or
    hormone), see CRF under
    hormones, specific
corticotropin-like intermediate-lobe
    peptide, see CLIP under hormones
    specific
cortisol (hydrocortisone) 85, 95, 96,
    236
craniopharyngioma, 229–31
CRF, see under hormones, specific

criss-cross inheritance, 126
Crohn's disease, 235
cross-sectional study, 10–13
curves, gr.: cross-sectional, 10
    of different tissues and body parts
        16–18
    distance and velocity, 6–7
    for foetal length, 38
    fractured, during stress, 103
    genetical control of, 128, 165
    gr. disorder, and, 226
    longitudinal, 10
    restriction of gr., following, 165–171
cyclic AMP, 84
cytoplasm, 28–9, 38, 84–5, 123
cytosine, 27

dendrites, 106, 108, 109, 111, 113, 115,
    118
diabetes, 100
diet, difference of between social
    classes, 151
dimorphism, sexual, see sex differences
disease: effects of, on gr., 147
    and secular trend, 161
    see also infection
DNA, 27–8, 30, 84, 85, 120
    and recombinant technology, 89, 92,
        232
Down syndrome, 37, 54, 174, 225
dynamic state of body constituents,
    25–6
dyslexia, 114

eczema, 95
ejaculation, 62
embryo, 36
    sex differentiation in, 51–5
emotional development, 80–3
    see also stress
encephalogram, see brain
endocrine glands, 55, 84
    and puberty, 100–3
endoplasmic reticulum, 28
energy: basal requirement, 26
    intake reduction during infection,
        133–4
    requirement in infancy, 131
evironment, effect on gr., 9, 119–64
    uterine, 40–3, 123
enzyme systems, effect on of birth, 49
epiphyses, 32, 33, 34, 76
evolution and: earlier maturation of
    girls, 56–7

fat metabolism, 90
gap between weaning and puberty,
    21–3, 102
hormones, 85, 88
racial differences in body shape,
    144–5
ritualized sex differentiation, 70
sex differences in finger length, 55
exercise: and GH, 90, 91
and social class, 151
eye: adolescent gr. spurt of, and
    myopia, 17
co-ordinated gr. of, 174
see also visual development

face, 17, 35, 67–8, 174, 228, 239
family resemblance, 123–6, 128,
    179–81, 211–13
fat: cells, structure of, 30
gr. curve, 17–18, 72–3, 96
increase in, due to cortisol
    treatment, 95
mechanism of growth of, 30, 31
metabolism of, and GH, 90–1
prenatal development of, 40
feet, gr. of, 67
fertilization, 36, 51, 99
foetus, 36
genetic programming of
    development of, 169–70
maldevelopment of, and later gr.
    disorder, 227–9
sensitive periods in development of,
    172, 173
thalidomide and, 172
follicle-stimulating hormones, see FSH
    under hormones, specific
FSH, see under hormones, specific

gene, 27, 84
dominant, 126
for height, and secular trend, 162
major, 124
poly, 124
pool, 140, 142
sex-limited and sex-linked, 126
see also heterosis
genetic instruction, 27
genetic programming, 165–174
genitalia, 54, 58–65, 97
development standards for, 215
genotype, 119, 120, 122–3, 127
germ cell, 27
gestation, length of, 37–8, 43–7

GH, see under hormones, specific
GHRH, see under hormones, specific
gigantism, 237–9
glandular tissue, regenerative powers
    of, 24–5
see also adrenal, endocrine,
    pituitary, thyroid glands
glucocorticoids, 95
GnRH, see under hormones, specific
gonad, 54
see also testis, ovary
gonadal dysgenesis, see Turner
    syndrome
gonadostat, 102
gonadotrophic hormones, 98, 101
insufficiency of, 231, 232
gonadotropin-releasing hormone, see
    GnRH under hormones, specific
Greulich–Pyle bone age standard, 76,
    78, 218, 221
growth: cell and organ, 28–30
cessation of, 19, 33–4, 157
compensatory, 165–6
delayed, 224–5
disease, minor, effects on, 147
disorders of, 229–39
hormones, and, 84, 85–6, 87–103
infection, effects on, 133–4
normal, standards of, 178–221
nutrition, effects on, 129–33
slow, due to cortisol treatment, 95;
    diabetes, 100; thyroid deficiency,
    93
stress, effects on, 147–8, 236–7
surveillance, 163–4
see also catch-up gr.
growth hormone, see GH under
    hormones, specific
growth hormone-releasing hormone,
    see GHRH under hormones, specific
growth-plate, 32–5
growth spurt, 6–7, 57, 66–8, 91, 96, 97,
    98
guanine, 27
gynaecomastia, 62

haemoglobin, foetal and adult forms,
    49, 74
hair: axillary, 62, 96, 97
body, 62
facial, 62, 97
pubic, 58, 60–2, 63–5, 96, 97, 98
development standards, 216
evolutionary function of, 70

hands, gr. of, 67
  as indicators of skeletal maturity,
    76–8
head: adolescent gr. of, 67–8
  prenatal gr. of 47–8 104
hearing, 112–13
height: adults, prediction of, for child
    from parents', 179–81
  charts, 180–212
  as diagnostic agent for pathology,
    213
  h.s, individual's, at different ages,
    correlation between, 42
  IQ, correlation with, 81
  variation in individual's, during day,
    186
  see also stature
hemispheres of the brain, 113–14
heterochronism, 174, 176–7
heterosis, **162**
HGH, see under hormones, specific
hips, 65, 68–70
histogenesis, **38**, 171
hormones, **84**–103
  androgenic, 85, 95–6, 97
  feedback mechanism and, 88, 93, 94,
    102
  gonadotrophic, see CGT, FSH, LH,
    under specific h.s
  hypothalamic, **87–8**
  mineralocorticoid, **93**
  mode of action of, **84–9**
  molecula structure of, 84, 87, 88, 89
  pituitary, 54, 55–6, 85–6
  pulsatility of secretion of, 89, 91, 93,
    95, 98, 101
  self-priming, 85
  steroid, see steroid hormones
  specific: ACTH, 86, 89, 95, 96,
    (insufficiency), 231;
    aldosterone, 93, CGT 54, **86**;
    CLIP, **97**; cortisol, 85, 95, 96,
    236; CRF, **87**; FSH, **86**, 88, 89,
    98–9, 101–2; GH, 85, 86, 88,
    89–92, 95, 98, 103, 223, 239;
    (deficiency), 89, 231, (secretion,
    influences on), 90, 91, 97, 98,
    148, 236–7, (treatment with),
    167–9, 224–5, 227–34; GHRH,
    88, 91; GnRH, 88, 96, 101, 102;
    HGH, 231, **232**; IGF–1, **33**;
    inhibin, 99; insulin, 85, 89, **100**,
    232; LH, **86**, 88, 89, 98–9, 101–
    2; LHRH, **88**; melatonin 85, 99–

100; oestradiol, 85, 102;
    oestrogen, 68, 88, 91, 97, **98**,
    101, 102, 225; placental
    lactogen, 86; progesterone, 99;
    prolactin **86** 88 **99**·
    somatomedin, 26, 34, 95;
    somatomedin–c, 33, **92**, 98;
    somatostatin, **88**; testosterone,
    54, 56, 68, 85, 91, 95, 96, 97–8,
    99, 101, 118, (effects on muscle
    fibres), 31, 40, 73, (in treatment
    of gr. delay), 224–5; thyroxine,
    85, 88, **93**, 169, 231, 232–3;
    TRH, **87**, 88, 93; TSG, 232,
    233; TSH, **86**, **87**, 88, 93, 231;
    vasopressin, 86
Human Growth Foundation (USA),
  222
human growth hormone, see HGH
  under hormones, specific
hydrocortisone, see cortisol
hypochondroplasia, **234–5**
hypothalamus, 55, 85, **86–8**, 91, 95, 97,
  101, 170, 232
hypothyroidism, 93, 169, 232–3

IGF–1, see under hormones, specific
infection: effect of, on gr., 133–4
  in pregnant women, 173
  period of peak vulnerability to, 131
infertility, chromosomal, 52, 227
  relative, in early adulthood, 62, 64
inhibin, see under hormones, specific
insulin, see under hormones, specific
insulin-like growth factor 1, see IGF–
  1 under hormones, specific
intellectual development: and
    lateralization of the brain, 114
  and number of siblings, 81
  and physical development, 80–1, 113
  and sensitive periods in brain
    development, 173
  and undernutrition, 131

jaw, 35

karyotype, **51**, 227
kidney: cells, regenerative powers of,
    24–5
  disease, side effects of corticoids in
    treatment of, 95
Klinefelter syndrome, 52
knee, 78
knemometer, 8, 95

Langerhans, islets of, 85, **100**
left-handedness, 114
leg: gr. of bones of, 34–5
   length, 66–7, 68, 74, 159
   and maturity gradient, 175–6
   operations to lengthen, 233
Leydig cells, 54, 85, 98, 99
Leydig-cell-stimulating hormone, *see*
   LH *under* hormones, specific
LH, *see under* hormones, specific
LHRH, *see under* hormones, specific
Little People of America (LPA), 222,
   233, 235
liver, regenerative powers of, 25
longitudinal record, 6–7, **10**, 178
   mixed, **10–13**
low birthweight babies **43–7**
luteinizing hormone, *see* LH *under*
   hormones, specific
luteinizing-releasing hormone, *see*
   LHRH *under* hormones, specific
lymphoid tissue, gr. curve, 17

maturation: of cerebral cortex, 109–13
   early and late, 103
   earlier, and secular trend, 159–60
maturity: biological, of foetus, 50
   dental, and thyroid deficiency, 93
   gradients, 175–6
   racial differences in, 144
measuring, 178, 179
   errors, 7–8
   of hormone secretions, 86, 89
   techniques, 182–6, 232
melatonin, *see under* hormones,
   specific
mebrane, cell, 26, 84
menarche, **64**, 102
   and breast development, 103
   and chronological and skeletal ages,
      79, 103
   family correlations in age of, 128
   geographical differences in, 145–6
   racial differences in, 142
   and secular trend, 159–62
   effects of urbanization on, 149
menstrual cycle, 55, 64, 88, 89, **98–99**
   anovulatory, 102
mental subnormality/handicap and
   'brain injury', 52, 53
   and thyroid deficiency in newborn
      babies, 93
milestones in development, 57, 79
mini-spurt, 8

mitosis, 28–30, 169–70
mongolism, *see* Down syndrome
monitoring of growth rates in infants,
   133
morphogenesis, **38**, 170
mosaic, **52**, 227
motor development, 50, 109–12, 176
   racial differences in, 144
moulding, 35
muscle: and androgen levels, 96
   exercise, effect on, 31
   fibres, **29**
   GH, effect on, 91
   gr. of, in puberty, 65, 67, 68, 70–2,
      97
   mode of action of, 24
   regenerative powers of, 24–5
   strength, 73–4
mutation, gene, **21**
myelin sheath, 107, 109, 111, 112–13

nerve cells (neurons), **106–11**, 115–18
   division of, 29, 38
   hormones, secretion of, by, 87
   prenatal development of, 93
   regenerative powers of, 25
nervous system, **84**, 107
   defects attributed to birth injury, 49
   development of, in preterm babies,
      50
   formation of connexions in, 115–18
   specialization of cells in, 171–2
neuroglia **29**, 38, **106–8**
neuron, **29**, **106–9**, 113, 116, 118, 172
nitrogen: balance during periods of
      infection, 133–4
   turnover, 25
nutrition: and cellular clock, 30–1
   and overfeeding of babies, 31, 40
   and undernourishment: adult and
      outcome following: u. in
      childhood, 134–6; u. in foetal
      life, 135–6, 169–70
   continuous in childhood, 122
   effects on gr., 129–33
   in pregnancy, 43
   and secular trend, 161

obesity, 31
   and early menarche, 83
   and social class, 151
   and steroid drugs, 95
oestradiol, *see under* hormones,
   specific

oestrogen, *see under* hormones,
  specific
orchidometer, 58, 63, 217
orthodontics, 35
orthopaedics, 35
ovarian follicles, 63, 102
ovary, 85
  degeneration of, 52, 54, 55
  development of, 63
  failure to develop, 227
ovulation, 64, 98, 102
ovum, 51, 98–9, 123
oxygen, lack of, at birth, 49

pancreas, 85, 100
  regenerative powers of, 24–5
pelvis: in girls at puberty, 68–70, 98
  sex difference in, at birth, 68
penis, 58–63, 97, 98
peptide, **84**, 87, 88, 92, 97
percentile, *see* centile
perinatal death, 45
  incidence of, related to gestational
    age, 46
'phase-differences', **11**
pineal gland, 85, 99–100
pituitary gland, 54, 55–6, 84, 85–9, 91,
  94, 95, 97, 98, 101–2
  and short stature, 229–32
  tumours of, 238–9
placenta, 36, 42
  failure to develop, 52
  hormonal activity of, 97
  tests for function of, 47
placental lactogen, *see under* hormones
  specific
plasticity, 35, 174
positive emission tomography (PET),
  113
post-term, **43**
poverty index, 151
pregnancy,
  drugs in, 172
  hormones in, 97
  infection in, 173
  'premature' babies, 43–7
  EEG of, 50
pre-term, **43**
primates: adrenal cortex of, 97
  brain gr. in, 137–9
  gr. curve and evolution, 21–3
  nutritional studies in, 137–40
  puberty in, and pineal gland, 100
  sensitive periods for visual function

in, 173
progesterone, *see under* hormones,
  specific
prolactin, *see under* hormones, specific
proportions, body: differences in,
  (racial) 142–4, (sex) 55, 57,
  (social) 131
  and secular trend, 159
  effects of undernutrition on, 130–1
prostate gland, 58, 62, 97
protein: high, diet, 25, 92
  low, diet, 25
  requirement in infancy, 131
  and RNA, 28, 84
  structure of, 28, 84
  synthesis, 170
  turnover, 25, 115
puberty, **58–74**, 98, 100–103
  in animals with breeding seasons,
    58, 100
  delayed, treatment of, 102
  hypothalamus and initiation of, 88
  induction of, to restrain gr., 237
  effects of melatonin on, 100
  precocious, 96, 174–5, 229
  rating-scales, 62
  standards, 213–18
  effects of testosterone at, 97–8
pulsatility of hormone secretion *see*
  under hormones

racial differences in size, shape and
  tempo of gr., 140–4, 178
radiography, *see* X-ray
receptors, 26, 34, **84**, 85, 87, 172
  down-regulated, 85
reflex, conditioned, 50
releasers, hypothalamic, (or hormone-
  releasing factors), **87–8**, 232
rheumatoid arthritis, 95
ribosomes, 28
ritualization, **70**
RNA, 27–8
rodents: brain gr. of, 136–7
  and catch-up gr., 170–1
  cerebellar development in, 118
  GH secretion of, 91
  gr., curve of, 21–2
  pituitary role in puberty of, 101
rubella (German measles) in
  pregnancy, 173
  and small-for-dates babies, 46

screening for: height as a proxy for

health, 162–4, 178–9
PKU, 93
small-for-dates babies, 47
thyroid deficiency in babies,
162–4, 178–9
scrotum, 58, 63
seasonal variation in gr., 8, 145–7
secular trend, 131, 156–62
seminal: vesicles, 58, 62, 97
emissions, 175
seminiferous tubules, 99
sensitive period, 172–5
and brain differentiation, 56
for cerebellar development in rat,
118
in development of foetus, 172, 173
sensor, hypothalamic, **88**
sex differences: adolescent gr. spurt,
in, 97, 100
in birthweight, 15, 57
in body proportions, 57, 65–70
in brain differentiation, 55, 56, 97
determination of, by chromosomes,
51–4
in relative length of fingers, 55
and gr. curves, 15, 59–65, 171
in GH secretion, 91
in heart, lungs and circulation, 74
in prepubertal brain function, 114
in rate of maturation, 15, 19, 56–7,
64–5, 78, 176, 219
in shoulder–hip ratios, 68–70, 100
and social function, 65, 70
sex, female, the 'basic', 55
sex-limited gene, *see under* gene
sex-linked gene, *see under* gene
sexual pathology, 56
shape, body, 122, 124
effects of undernutrition on, 130–1
shortness, *see* stature
shoulder: gr. of at puberty, 65, 68–70,
98
s.–hip-width dimorphism, 68–70
sibs, effects of having, on gr., 151–2
Silver–Russell syndrome, 228
skeletal maturity, 76–8
and adrenal androgens, 96
delay (due to cortisol treatment), 95
(due to thyroid deficiency), 93
family correlations of, 128
racial differences in, 144
skull, 35, 67, 174
foetal development, 104
sleep, secretion of GH during, 91

'small-for-dates' babies, 43–7, 227–9
treatment with GH, 228
smoking in pregnancy, small-for-dates
babies, and, 47, 147
social class: and diet, 151
obliteration of s.c. height
differences, 154–6
social mobility, 81
and height of children, 153–4
social status and physical maturity, 83
socioeconomic factors, effects on size
and tempo of gr., 149–56
somatomedin, *see under* hormones,
specific
somatomedin–c, *see under* hormones,
specific
somatostatin, *see under* hormones,
specific
sperm, 5, 99
spinal cord, 105, 106, 117
in rat, 118
spline, **9**
sport, 144
starvation, effects of on gr., 129
and GH, 90–1, 92
stature, 9
and emotional and intellectual
development, 80
genetical control of, 124, 126–8
and GH, 91–2
and IGF-1
maternal, effect on foetus, 41–3
and mental ability, 156
and pubertal development, 64–5
short, 222–37
causes of, (absorption of food,
disorders of) 235, (cartilage and
bone disorders) 233–5,
(chromosomal disorders) 225–7,
(diseases of heart, lung and
kidney) 236, (endocrine disorders)
229–33, (genetic) 223–4, (gr.
delay) 224–5, (intrauterine
maldevelopment) 227–9,
(psychological) 236–7, (steroids
and radiation) 236
and sibs, number of, 81
and social class, 134
social class differences in,
obliteration of, 154–6
tall, 237
related to chromosomal
abnormality, 52
see also height

steroid hormones, 85
  anabolic, in treatment of:
    delayed growth, 225
    Turner syndrome patients, 227
  therapeutic, as cause of stunted gr
    236
Still's disease, see rheumatoid arthritis
strength, 73–4
  'outgrowing' of, 74
stress, effects on gr., 148, 236–7
supine length, 15
synapse, 113, 115, 117

tallness, see stature
Tanner–Whitehouse bone-scoring,
  76–8, 218–21
teeth, 35, 57, 78–9
  canine, in primates, 68
  eruptions of, family correlations, 128
  see also, age, dental, and maturity,
    dental
tempo of gr., 75, 80, 103, 122, 128,
  224–5
  genetical control of, 128
  malnutrition, effect on, 129–33
  racial differences in, 142, 144
  and socioeconomic factors, 149–56
  urbanization, effects on, 148
testis, 54, 55, 58–63, 101
  development of, in embryo, 54
  at puberty, 215, 217–18
testosterone, see under hormones,
  specific
thalidomide, 172
thymidine, 27
thyroid: gland, 85,
  hormone, see thyroxine, under
    hormones, specific
thyroid-stimulating hormone
  (thyrotrophin) see TSH under
  hormones, specific
thyrotrophin-releasing hormone, see
  TRH under hormones, specific
thyroxine, see under hormones,
  specific
TRH, see under hormones, specific

triple, 21, 54
triplet code, 28
trunk length, 66–7, 74
Turner syndrome, 52
  treatment with GH, 227
twins: prenatal growth, 40, 122
  mono- and dizygotic, effects of
    environment on, 121–3

ultrasound, 36, 42, 47, 104
  small-for-dates babies, and, 47
  development of testes, and, 58
  development of uterus and ovaries,
    and, 63
unemployment, of father and height of
  children, 152–3
urbanization, effects of on gr., 148–9
uterine environment, effects of on
  development of foetus, 40–3
uterus, 58, 63, 64, 85, 98, 99

vagina, 98
variation in gr. rate, 8
vasopressin, see under hormones,
  specific
vertebral column, 97
visual development, 108, 109, 111–17,
  171–2
  binocular, 115, 120
  and gr. of eye, 17, 174
voice breaking, 62

weight charts, 196–9, 201, 203, 206,
  207, 208, 209
wrist, 35, 76–8, 218

X–ray: cerebellar development in rat
  halted by, 118
  in differential diagnosis of
    achondroplasia and
    hypochondroplasia, 235
  GH deficiency, diagnosis of, by, 229
  skeletal maturity, assessment of, by,
    76–8, 218
  small stature, as cause of, 231, 236